gateway science

OCR
Science
for GCSE WITHDRAWN

Byron Dawson
Bob McDuell
Mike Brimicombe

Series editor: Bob McDuell

www.heinemann.co.uk

✓ Free online support
✓ Useful weblinks
✓ 24 hour online ordering

01865 888058

Heinemann

Inspiring generations

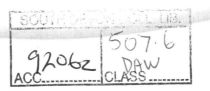

Heinemann Educational Publishers
Halley Court, Jordan Hill, Oxford OX2 8EJ
Part of Harcourt Education

Heinemann is the registered trademark of Harcourt Education
Limited

First published 2006

10 09 08 07 06
10 9 8 7 6 5 4 3 2 1

10-digit ISBN: 0 435 67522 2
13-digit ISBN: 978 0 435 67522 6

Edited by (TBA) Bob McDuell
Designed by Wooden Ark
Typeset by HL Studios, Long Hanborough, Oxford
Copyedited by Marilyn Grant and Felicity Kendall

Original illustrations © Harcourt Education Limited, 2006

Illustrated by HL Studios
Printed in the UK at Bath Colourbooks
Cover photo: Getty Images©
Picture research by Ginny Stroud-Lewis

Acknowledgements
The authors and publisher would like to thank the following
individuals and organisations for permission to reproduce
photograph and artwork:

Page 2, Getty Images; 3, Ian Hooton / SPL; 5, Michael Donne / SPL (x3); 5, **BR** Aaron
Haupt / SPL; 6, Alamy Images / Sally and Richard Greenhill; 7, Getty Images / Food
Pix; 8, Andy Crump, TDR, WHO / SPL; 10, Corbis / Brooke Fasani; 11 Alamy Images
/ David Hoffman Photo Library; 12, **T** Martin Dohrn / SPL; 12, **B** Corbis / Sygma /
Rogate Joe; 13, SPL / BSIP, LA; 14, Corbis / Greg Smith; 15, Getty Images / PhotoDisc;
18, **T** Author supplied; 18, **B** Rory McClenaghan / SPL; 19, **T** Harcourt Education
/ Ginny Stroud-Lewis; 19, **B** SPL / Tek Image; 20, Wellcome Medical Photographic
Library; 22, Alamy Images / Alex Segre; 23, Getty Images / PhotoDisc (x2); 26, Corbis;
27, Getty Images / PhotoDisc; 28, Author supplied; 29, SPL / Biophoto Associate; 30,
www.genome.gov; 31, Alamy Images / Janine Wiedel Photolibrary; 31, **B** Dr Jeremy
Burgess / SPL; 33, **T** Alamy Images; 33, Dept. Of Clinical Cytogenetics, Addenbrookes
Hospital / SPL (x2); 34, SPL / Simon Fraser / RVI, Newcastle-Upon-Tyne; 38, The
Kobal Collection / Amblin / Universal; 39, **T** Image Quest; 39, **B** Alamy Images / Glyn
Thomas; 41, **T** Education Photos / Alamy; 41, **B** UK Ladybird Survey / Peter Brown; 42,
Alamy; 43, **L** Museum Images / Pierre Fidenci; 43, **R** Getty Images / Photodisc; 44, **T**
SPL; 44, **M** Alamy; 44, **B** Getty Images / Photodisc; 45, **T** Getty Images / Photodisc;
45, **B** Corbis; 46, Alamy / Royalty Free; 47, OSF; 48, **T** Paola Zucchi / ABPL; 48, **B**
Getty Images / Photodisc; 50, Getty Images / PhotoDisc; 51,

T Alamy Images / Motoring Picture Library; 51, **B** Corbis / Juan Medina / Reuters; 52,
T The Photolibrary Wales / Alamy; 52, **B** Corbis; 53, **TR** K.H. Kjeldsen / SPL; 53, **TL**
Mediscan; 53, **B** ImageState / Alamy; 54, Alamy Images / Andrew Darrington; 55, **B**
Corbis / Chris Mattison; Frank Lane Picture Agency; 55, **T** Getty Images / PhotoDisc;
56, **T** Corbis; 56, **B** NHPA / Mike Lane; 57, Getty Images / Photodisc; 58, SPL / B.
Murton / Southampton Oceanography Centre; 59, **T** Getty Images / Botanica / Emily
Brooke Sandor; 59, B SPL / Sinclair Stammers; 60, **T** Corbis / Bettmann; 60, **B** OSF /
Peter Parks; 61, **T** OSF / Peter Parks; 61, **B** OSF / David Fox; 62 Getty Images / Hulton
Archive; 63, **T** Alamy Images / PCL; 63, **B** Alamy Images / David R. Frazier Photolibrary,
Inc.; 64 NASA / SPL; 65, **T** Michael Donne / SPL; 65, **ML** Adrian Davies / naturepl.com;
65, **MR** Ken Preston-Mafham / premaphotos.com; 65, **B** Dr Jeremy Burgess / SPL;
66, **T** SPL / Dr Jeremy Burgess; 66, **B** Natural Visions / Heather Angel; 67, **M** Roger
Key / English Nature; 67, **B** Mark Carwardine / naturepl.com; 67, **T** Doug Perrine /
naturepl.com; 68, **T** Paul Glendell / Alamy; 68, **B** NHPA; 69, NHPA / Trevor McDonald;
70, **T** Corbis / Amos Nachoum; 70, **B** f1 online / Alamy; 74, Alamy Images / Robert
Harding Picture Library Ltd / John Miller; 75, **T** Corbis; 75, **B** Harcourt Education /
Ginny Stroud-Lewis; 76, Maximilian Stock Ltd/ABPL (x2); 77, Tim Hill / ABPL; 78,
Getty Images / Taxi; 79, **T** Harcourt Education / Ginny Stroud-Lewis; 79, **B** Harcourt
Education; 80, Alamy; 81, Harcourt Education / Ginny Stroud-Lewis; 82, Alamy Images
/ David Young-Wolff; 83, **T** Corbis; 83, **B** Alamy Images / Travel-Shots; 85, Harvey
Pincis / SPL; 86, Digital Vision; 87, Corbis; 89, Simon Fraser / SPL; 90, Corbis / John
Gress / Reuters; 91, **T** Motoring Picture Library / Alamy; 91, **B** SPL; 93, Harcourt
Education / Peter Gould; 94, Science and Society Picture Library; 95, David Hoffman
Photo Library / Alamy; 97, **L** Digital Vision; 97, **R** Andrew Lambert Photography / SPL;
98 Corbis / Chinch Gryniewicz; Ecoscene; 99, **T** Alamy Images / David Lyons; 99, **M**
Alamy Images / Paul Glendell; 99, **B** Topfoto / UPPA; 101, **T** Charles D. Winters / SPL;
101, **B** David Taylor / SPL; 102 Alamy Images / By Ian Miles-Flashpoint Pictures; 103,
Creatas; 106, Getty Images / Photonica / Johner; 110, Rex / CNP; 111, Alamy Images /
Ron Scott; 113, Klaus Guldbrandsen / SPL; 114, Alamy Images / Mitch Diamond; 115,
T Corbis / Simon Kwong / Reuters; 115, **B** Harcourt Education / Ginny Stroud-Lewis;
116, **T** Simon Fraser / SPL; 116, **M** Sheila Terry / SPL; 116, **B** Dirk Wiersma / SPL;
117, KPT Power Photos; 118, Digital Vision; 119, **T** Adam G. Sylvester / SPL; 119, **B**
Corbis / Mohsin Raza / Reuters; 121, **T** Corbis / Layne Kennedy; 121, **B** SPL; 123, SPL
/ Andrew Lambert Photography; 124, Maximilian Stock Ltd / SPL; 126, Department
of Geological & Mining Engineering & Sciences, Michigan Technological University;
127, Alamy Images / Photo Japan; 129, Alamy Images / Motoring Picture Library; 130,
Iain Farley / Alamy; 131, SPL / BSIP, M.I.G. / BAEZA; 134, Alamy Images / FLPA; 135,
T Alamy Images / foodfolio; 135, **B** Alamy Images / Marie-Louise Avery; 138, **T** SPL /
Jonathan Watts; 138, **B** SPL / Mehau Kulyk & Victor De Schwanberg; 139, **T** ?; 139,
B Corbis / Sygma / Petit Claude; 140, **T** Harcourt Education Ltd / Peter Gould; 140,
B SPL / Malcolm Fielding, Johnson Matthey PLC; 141, Harcourt Education Ltd / Peter
Gould; 142, SPL / Charles Bach; 146, Getty Images / Photodisc; 147, **T** Alamy Images
/ Photofusion Picture Library / Christa Stadtler; 147, **B** Dr. Arthur Tucker / SPL; 150,
Corbis; 154, Getty Images; 155, **T** Getty Images / Iconica / Frazer Cunningham; 155,
B Corbis / Tim Graham; 158, Getty Images / PhotoDisc; 159, **T** Alamy Images / Alan
Oliver; 159, **B** Alamy Images / Jim Wileman; 161, **T** Alamy Images / Adrian Sherratt;
161, **B** BennettPhoto / Alamy; 162, **B** Getty Images / PhotoDisc; 162, **T** Alamy/Royalty
Free; 163, **T** Corbis / Lucidio Studio, Inc.; 163, **B** Alamy / Royalty Free; 164, Getty
Images / Iconica / Andre Cezar; 166, SPL / Tek Image; 167, Corbis; 168, SPL / NASA;
170, Alamy Images / Joe Tree; 171, Corbis; 174, **T** Corbis; 174, **B** SPL / NASA; 175,
T Alamy Images / Photofusion Picture Library / Brenda Prince; 175, B SPL / Dr P.
Marazzi; 177, Corbis / Reuters; 178, Corbis / Reuters; 182, Getty Images / Photodisc;
183, **B** Empics / AP; 183, **T** Corbis; 184, Getty Images / Photodisc; 186, Alamy Images
/ Robert Harding Picture Library Ltd; 187, Martyn F. Chillmaid / SPL; 190, Corbis /
Bettmann; 191, Roger Ressmeyer / Corbis; 192, **T** Paul Avis / SPL; 192, **B** Getty Images
/ Photodisc; 194, SPL / James King-Holmes; 195, **T** Corbis / Bettmann; 195, **B** Getty
Images; 197, Medical-on-Line / Alamy; 198, Alamy Images / Mark Boulton; 199,
Getty Images / PhotoDisc; 200, Corbis; 201, **R** Phototake Inc. / Alamy; 201, **L** Corbis
/ Sygma / Tiziou Jacques; 202, SPL / European Space Agency; 203, **T** David
Nunuk; 203, **B** SPL / European Space Agency; 205, NASA / SPL; 206, Getty Images /
PhotoDisc; 207, T Getty Images / PhotoDisc; 207, **B** Corbis; 208, **R** SPL / Roger Harris;
208, **L** Getty Images / The Image Bank / Peter Lilja; 209, NASA / ESA / STScI / SPL;
210, Novosti Press Agency; 211, **B** National Optical Astronomy Observatories
/ SPL; 212, Roger Ressmeyer / Corbis; 214, SPL / Emilio Segre Visual Archives /
American Institute of Physics; 49 and 72, www.bbc.co.uk/gcsebitesize; 96, W.L. Gore
and Associates. Copyright©.

Introduction

This student book covers the higher tiers of the new OCR Gateway Science specification. The first examinations are in January 2007. It has been written to support you as you study for the OCR Gateway Science GCSE.

This book has been written by examiners who are also teachers and who have been involved in the development of the new specification. It is supported by other material produced by Heinemann, including online teacher resource sheets and interactive learning software with exciting video clips, games and activities.

Part of this new GCSE is the type of assessments of your work that will be carried out by your school. These are called 'Science in the News' and 'Can Do Tasks' and are explained fully on pages 218–221.

We hope this book will help you achieve the best you can in your GCSE core Science Award and help you understand how much science affects our everyday lives. As citizens of the 21st century you need to be informed about science issues such as nuclear power, genetic engineering and pollution. Then you can read newspapers or watch television programmes and really have views about things that affect you and your family.

For most of you we hope that you will go on to do Additional Science which will be a second Science GCSE. During this you will cover the rest of the Science you will need to be able to go on and study AS courses in Physics, Chemistry or Biology.

The next two pages explain the special features we have included in this book to help you to learn and understand the science, and to be able to use it in context. At the back of the book you will also find some useful tables, as well as a glossary and index.

About this book

This student book has been designed to make learning science fun. The book follows the layout of the OCR Gateway specification. It is divided into six sections that match the six modules in the specification with two for Biology, two for Chemistry and two for Physics: B1, B2, C1, C2, P1, P2.

The module introduction page at the start of a module (eg.below right) introduces what you are going to learn. It has some short introductory paragraphs, plus 'talking heads' with speech bubbles that raise questions about what is going to be covered.

Each module is then broken down into eight separate items (a–h), for example, B1a, B1b, B1c, B1d, B1e, B1f, B1g, B1h.

Each 'item' is covered in four book pages. These four pages are split into three pages covering the science content relevant to the item plus a 'Context' page which places the science content just covered into context, either by news-related articles or data tasks, or by examples of scientists at work, science in everyday life or science in the news.

Throughout these four pages there are clear explanations with diagrams and photos to illustrate the science being discussed. At the end of each module there are three pages of questions to test your knowledge and understanding of the module.

There are three pages of exam-style end of module questions for each module.

The talking heads on the Module intro page raise questions about what you are going to learn.

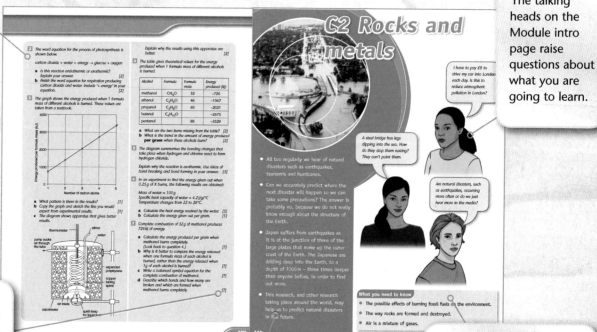

The numbers in square brackets give the marks for the question or part of the question.

The bulleted text introduces the module

This box highlights what you need to know before you start the module.

Context pages link the science learnt in the item with real life.

This box highlights what you will be learning about in this item.

General approach to the topic

Question box at the end.

Clear diagrams to explain the science.

Questions in the text make sure you have understood what you have just read.

Some amazing facts have been included – science isn't just boring facts!

When a new word appears for the first time in the text, it will appear in bold type. All words in bold are listed with their meanings in the glossary at the back of the book.

The keywords box lists all keywords in the item.

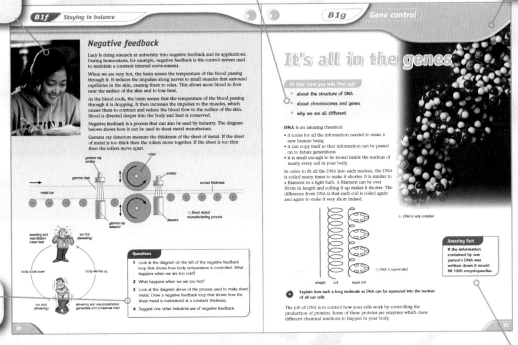

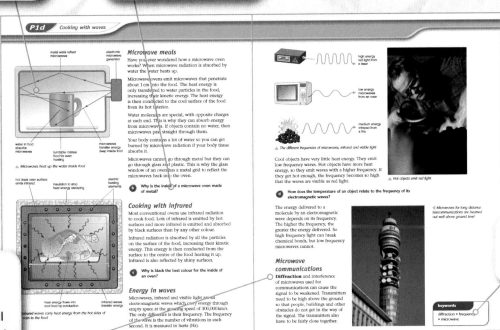

Contents

B1 Understanding ourselves   2

B1a   Fit for life   3
B1b   What's for lunch?   7
B1c   Keeping healthy   11
B1d   Keeping in touch   15
B1e   Drugs and you   19
B1f   Staying in balance   23
B1g   Gene control   27
B1h   Who am I?   31
End of module questions   35

B2 Understanding our environment   38

B2a   Ecology in our school ground   39
B2b   Grouping organisms   43
B2c   The food factory   47
B2d   Compete or die   51
B2e   Adapt to fit   55
B2f   Survival of the fittest   59
B2g   Population out of control?   63
B2h   Sustainability   67
End of module questions   71

C1 Carbon chemistry   74

C1a   Cooking   75
C2b   Food additives   79
C2c   Smells   83
C2d   Making crude oil useful   87
C2e   Making polymers   91
C2f   Designer polymers   95
C2g   Using carbon fuels   99
C2h   Energy   103
End of module questions   107

C2 Rocks and minerals — 110

C2a Paints and pigments — 111
C2b Construction materials — 115
C2c Does the Earth move? — 119
C2d Metals and alloys — 123
C2e Cars for scrap — 127
C2f Clean air — 131
C2g Faster or slower (1) — 135
C2h Faster or slower (2) — 139
End of module questions — 143

P1 Energy for the home — 146

P1a Heating houses — 147
P1b Keeping homes warm — 151
P1c How insulation works — 155
P1d Cooking with waves — 159
P1e Infrared signals — 163
P1f Wireless signals — 167
P1g Light — 171
P1h Stable Earth — 175
End of module questions — 179

P2 Living for the future — 182

P2a Collecting energy from the Sun — 183
P2b Power station (1) — 187
P2c Power station (2) — 191
P2d Nuclear radiations — 195
P2e Our magnetic field — 199
P2f Exploring our Solar System — 203
P2g Threats to Earth — 207
P2h The Big Bang — 211
End of module questions — 215

Useful Data — 218
Periodic Table — 219
About can-do tasks — 220
About science in the news — 222
Glossary — 224
Index — 230

B1 Understanding ourselves

This unit is about understanding ourselves. It is only by understanding who and what we are that we can learn to make the right decisions that will keep us healthy and help us to enjoy a long and happy life. In this module you will learn how we use food to obtain energy and what happens in parts of the world where food is scarce and people are starving. You will also learn how we are affected by disease and what we can do to keep healthy.

One of the reasons that humans have been so successful and are found living all over the world is that our bodies can regulate and keep constant many of our internal processes, such as body temperature. However, these internal mechanisms can be altered by drugs. Some people's lives have been ruined by drugs. You will learn about different types of drugs and the affect they can have on the body.

Finally, you will learn about DNA and how it makes us who and what we are. The study of DNA is an exciting new branch of science and it promises to bring about many new and wonderful changes in your lifetime.

I can't be bothered to get fit. I am healthy and you have got to die of something so I may as well enjoy life.

Being fit can be enjoyable. You feel so much better when you are healthy and fit.

What you need to know

- About types of food, what it contains and that it provides us with energy.

- How we can stay fit and healthy.

- Microbes can cause disease.

- Variation exists in animals and plants and that we are all different.

Fighting fit

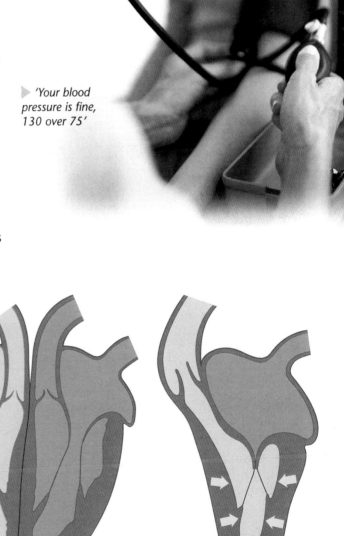

'Your blood pressure is fine, 130 over 75'

In this item you will find out

- about your blood pressure and what happens if it gets too high or too low

- about the difference between fitness and health and how fitness can be measured

- about aerobic and anaerobic respiration

Have you ever had your blood pressure taken by a doctor? When the doctor gives you the result you get two readings instead of one.

Your heart beats about 80 times every minute. When it beats, blood is forced through the blood vessels and your blood pressure is at its highest. This is called the **systolic** pressure and is the first number given.

When your heart is resting the narrow blood vessels slow the blood flowing through them, which lowers the pressure. This is called the **diastolic** pressure and is the second number given. Blood pressure is measured in units called mm Hg.

Your blood pressure changes all of the time. It depends on how active you are, your age, weight and lifestyle, how much alcohol you drink and whether you are angry or calm.

a Explain the difference between systolic and diastolic pressure.

Amazing fact

A young fit person may have a blood pressure of about 120 mm Hg over 70 mm Hg.

▲ Heart resting   ▲ Heart contracting

High and low pressure

If your blood pressure is too high or too low this can cause problems. High blood pressure can cause weak blood vessels to burst. If this happens in the brain, it can cause a 'stroke' which can damage your brain.

High blood pressure can also damage organs such as the kidneys. Some people have low blood pressure. This can lead to poor blood circulation, dizziness and fainting as the brain does not get enough oxygenated blood. The table shows normal blood pressure and high and low blood pressure.

Blood pressure	Systolic (mm Hg)	Diastolic (mm Hg)
normal	130	75
high	160	105
low	90	40

Aerobic respiration

Your body cells get energy by reacting glucose with oxygen. This is called **aerobic respiration**.

▶ *Aerobic respiration*

glucose + oxygen $\longrightarrow$ carbon dioxide + water + energy

$$C_6H_{12}O_6 + 6O_2 \rightarrow 6CO_2 + 6H_2O \quad (+ \text{energy})$$

Anaerobic respiration

When you do vigorous exercise your heart and lungs cannot provide your muscles with enough oxygen quickly enough. When this happens, your cells carry out **anaerobic respiration** as well as aerobic respiration.

▶ *Anaerobic respiration*

glucose $\longrightarrow$ lactic acid + energy

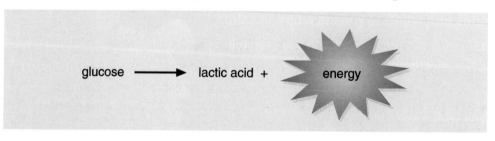

> **Examiner's tip**
>
> The equation for respiration is the same as the equation for photosynthesis in reverse.

As well as energy, **lactic acid** is also produced during anaerobic respiration due to the incomplete breakdown of glucose. This lactic acid can build up in your muscles and cause muscle fatigue and pain. Anaerobic respiration does not produce as much energy as aerobic respiration.

b Describe three differences between aerobic and anaerobic respiration.

c Suggest why our bodies do not use anaerobic respiration all of the time.

Recovering from fatigue

When you do hard exercise it induces a lack of oxygen in your cells. This is called the **oxygen debt** and it has to be repaid.

When you respire anaerobically, glucose is broken down into lactic acid instead of carbon dioxide and water. When you stop exercising you are usually panting. You continue to breathe heavily until your lungs have provided your body with enough oxygen to break down all of the lactic acid into carbon dioxide and water. The carbon dioxide is then carried in the blood to the lungs where you breathe it out.

Your heart rate is also increased and this helps your blood to carry lactic acid to your liver where it can be broken down.

▲ Breathing and heart rate increase to deliver glucose and oxygen to the muscles

Fitness

Fitness is not the same as being healthy and free from disease. Fitness is how efficiently your body can perform some of its functions. It is usually a result of exercising.

▲ This person is well but not fit

▲ This person is fit but not well

▲ Lactic acid builds up causing fatigue and pain

d   **The overweight man is well but not healthy. Suggest why.**

There are different ways of measuring fitness. You can measure how strong a person is by seeing how many press-ups they can do, and you can measure stamina by seeing how long they can keep doing an exercise. You can also measure agility, flexibility and speed.

---

keywords

aerobic respiration •
anaerobic respiration •
diastolic pressure • lactic
acid • oxygen debt •
systolic pressure

The hidden problem

Nearly one in three people has high blood pressure. About 30% of people with high blood pressure are not aware that they have it. It tends to be older people who have higher blood pressure.

High blood pressure is defined as being higher than 140 mm Hg over 90 mm Hg. If high blood pressure is diagnosed it can be successfully treated in most people. If it is left untreated it can cause a variety of problems.

The graph shows the incidence of high blood pressure in different groups of people.

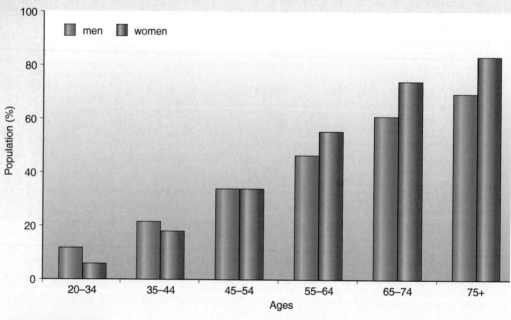

Questions

1   What proportion of the population has high blood pressure?

2   Does a person with a blood pressure of 140 mm Hg over 85 mm Hg have high blood pressure? Explain your answer.

3   Why is it important to diagnose high blood pressure early?

4   Which part of the population tends to have the highest blood pressure?

5   Which group of the population has shown the least change in the incidence of high blood pressure as they get old?

6   Average blood pressure fell between 1976 and 1994. Suggest why.

Diet dilemmas

ONLY A YEAR TO EAT ALL THIS!

In this item you will find out

- what is meant by a balanced diet and about protein intake

- about the digestive system

- how to calculate whether you are over or underweight

Amazing fact

Each year you eat about 500 kg of food. That's about the weight of 20 sacks of potatoes.

Do you eat a healthy balanced diet with lots of fruit and vegetables, meat, fish and whole grains?

Eating a balanced diet can be complicated because it can vary depending on how old you are, what gender you are and how active you are.

- young people who are growing need to eat more than older people who have stopped growing
- active people need to eat more than people who are not active
- men need to eat more than women.

There are also other factors that influence the types of food that people eat.

Vegetarians and vegans choose not to eat meat so they need a different kind of balanced diet from people who do eat meat.

People sometimes avoid certain foods for religious reasons and some people can have medical reasons for avoiding certain types of food, for example people who are allergic to peanuts.

a Why do you think that men need to eat more than women?

What's in our food?

Example of type of food	Food group	What we digest it into
Eggs	Protein	Amino acids
Butter	Fat	Fatty acid and glycerol
Potato	Carbohydrate	Simple sugars such as glucose

Animal proteins are called 'first class proteins' because they contain all the essential amino acids we need, but which we can't make for ourselves.

Most plant proteins only contain some of the amino acids that we need. Different plant proteins contain different amino acids.

b **Suggest how a vegetarian could get all of the amino acids that they need.**

We can calculate our recommended average daily protein intake by using the following formula:

RDA in g = 0.75 × body mass in kg

(RDA = recommended daily average)

c **Emma weighs 48 kg. How much protein should she eat each day?**

If we do not get enough protein in our diet, then we could suffer from protein deficiency (**kwashiorkor**). People in developing countries often suffer from it.

▶ *This child is suffering from a lack of protein in his diet*

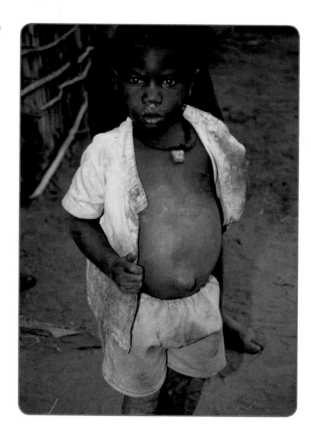

Digestion

When food is eaten, muscles in the gut contract and push the food along. As the food is pushed from the mouth to the anus, it is broken down by digestive **enzymes** into much smaller molecules. These small molecules can then be absorbed through the gut wall and into the blood plasma or the lymph to be used by the body. This is called **chemical digestion**.

 d Suggest why food can only be absorbed when it is broken down into small molecules.

Food group	Digestive enzyme
Protein	Proteases
Fat	Lipases
Carbohydrate	Carbohydrases

Different food groups are digested by different enzymes in our mouths, stomachs and small intestines.

Food is kept in the stomach for several hours. During this time, hydrochloric acid is added to the food. This kills most of the bacteria on the food and helps to break it up into smaller molecules. Acid in the stomach also helps the enzymes to work.

 e Why is it important to destroy most of the bacteria found on our food?

Food then passes into the small intestine where more enzymes are added to the food.

Bile is also added to the food in the small intestine. This improves fat digestion. Bile is not an enzyme. It is a chemical produced by the liver and stored in the gall bladder. When it is added to food, it makes the fat break up into smaller droplets. It works in the same way as washing up liquid breaks up fat when you wash greasy plates. This is called **emulsification**. This increases the surface area of the fat for the enzymes to work on.

 f Suggest why breaking fat up into smaller droplets makes it easier for enzymes to break down the fat.

The small molecules of food are then absorbed through the walls of the small intestine and into the blood plasma or the lymph. This process happens by **diffusion**.

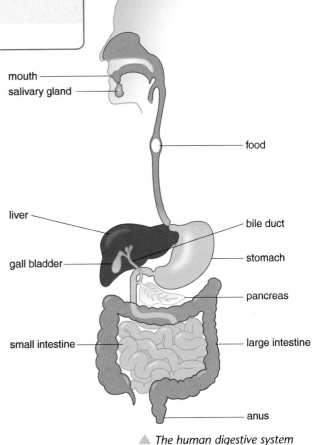

mouth
salivary gland
food
liver
bile duct
gall bladder
stomach
pancreas
small intestine
large intestine
anus

▲ *The human digestive system*

keywords

bile • carbohydrase • chemical digestion • enzyme • diffusion • emulsification • kwashiorkor • lipase • protease

Body image

Sarah is 15. She spends a lot of time reading fashion magazines and she is very depressed about how she looks. She has been teased by some of the girls at her school. She wishes that she could look like one of the slim and attractive models she sees in the magazines then perhaps she wouldn't be bullied at school. She has started to eat less food. At first her parents didn't notice but she has lost a lot of weight and they are becoming worried.

They take her to see the doctor. Dr Mackay examines Sarah and explains to her that depending on sex, age and height, we all have an ideal weight. Dr Mackay tells Sarah that dieting can lead to problems such as anorexia, and being very underweight can have severe health risks. She suggests that Sarah talks to a counsellor about her feelings.

Sarah can work out her ideal weight by using the following formula to calculate her body mass index (BMI).

$$\text{BMI} = \text{mass in kg}/(\text{height in m})^2$$

She can then compare her BMI with the table to see if she is at a normal weight or is underweight or overweight.

Mass (kg)

Height (cm)	54	59	64	68	73	77	82	86	91	95	100	104	109	113
137	29	31	34	36	39	41	43	46	48	51	53	56	58	60
142	27	29	31	34	36	38	40	43	45	47	49	52	54	56
147	25	27	29	31	34	36	38	40	42	44	46	58	50	52
152	23	25	27	29	31	33	35	37	39	41	43	45	47	49
158	23	24	26	27	29	31	33	35	37	38	40	42	44	46
163	21	22	24	26	28	29	31	33	34	36	38	40	41	43
168	19	21	23	24	26	27	29	31	32	34	36	37	39	40
173	18	20	21	23	24	26	27	29	30	32	34	35	37	38
178	17	19	20	22	23	24	26	27	29	30	32	33	35	36
183	16	18	19	20	22	23	24	26	27	28	30	31	33	34
188	16	17	18	19	21	22	23	24	26	27	28	30	31	32
193	15	16	17	18	20	21	22	23	24	26	27	38	29	30
198	14	15	16	17	19	20	21	22	23	24	25	27	28	29
203	13	14	15	17	18	19	20	21	22	23	24	25	26	28

☐ underweight   ☐ healthy weight   ☐ overweight   ☐ obese

Questions

1 Why does Sarah want to be thin?

2 Sarah weighs 54 kg and is 1.73 m tall. Calculate her BMI.

3 Look at the table. Is Sarah underweight, overweight or normal weight?

4 What may happen if Sarah loses too much weight?

5 Why did Dr Mackay suggest that Sarah sees a counsellor?

Fighting disease

In this item you will find out

- about different things that can make us ill

- how our body responds to illness

- how drugs are tested to make sure they are effective and safe

Illness can be caused by many different things.

Cancer occurs when body cells continue to divide uncontrollably. They produce a mass of cells called a **tumour**. Some tumours stop growing and are called **benign**. It is the tumours that continue to grow and spread that are dangerous. These are called **malignant**.

We can make changes to our lifestyles which may reduce the risk of getting some cancers. We can avoid too much sunshine, which contains UV light, smoking or certain types of food. If we eat lots of fresh fruit and vegetables containing antioxidants, this may reduce our chances of getting cancer.

The table shows the estimated survival rates after five years for men diagnosed with different types of cancer.

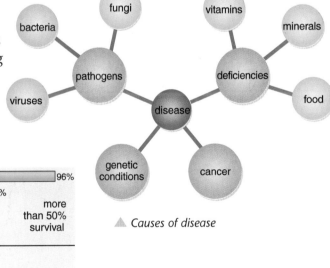

▲ *Causes of disease*

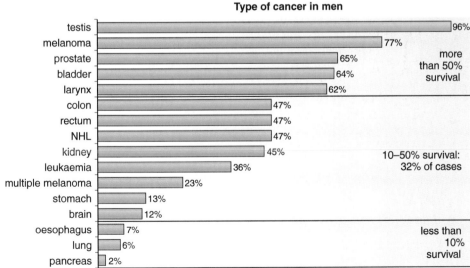

Type of cancer in men

Type	Survival
testis	96%
melanoma	77%
prostate	65%
bladder	64%
larynx	62%
colon	47%
rectum	47%
NHL	47%
kidney	45%
leukaemia	36%
multiple melanoma	23%
stomach	13%
brain	12%
oesophagus	7%
lung	6%
pancreas	2%

more than 50% survival

10–50% survival: 32% of cases

less than 10% survival

a Suggest which type of cancer is most easily treated in men.

b Suggest which type of cancer is the most difficult to treat in men.

Malaria

▲ *Mosquitoes spread malaria*

Malaria is one of the world's biggest killers. It is caused by a **parasite** that lives in the blood and liver. The human body and other animals act as a **host** to this parasite. The parasite is spread by a mosquito.

The mosquito sucks up the blood of an animal that has the parasite. The mosquito then carries the parasite and when it sucks the blood of a human, the parasite passes into the person's blood stream and causes malaria. Animals, such as the mosquito, that carry disease-causing organisms from one animal to another, are called **vectors**.

Controlling vectors

Diseases, such as malaria, can be prevented by destroying the mosquito vectors. Mosquitoes breed in water such as ponds and even puddles.

They can be killed by spraying them with chemicals, or by putting fish into the ponds that eat the mosquito larvae. People can take drugs to kill the parasite in their bodies, or use nets over their beds so that the mosquitoes cannot bite them during the night.

c Suggest how other vectors such as the housefly could be controlled.

Antibodies and antigens

Microorganisms cause disease when they damage the cells in our bodies or produce poisonous chemicals called **toxins**. Microorganisms that do this are called **pathogens** and each pathogen has its own **antigens**.

Our bodies respond by producing different **antibodies** for each antigen. The antibody locks onto the antigen and kills the pathogen.

▲ *Mosquitoes can be killed by chemicals*

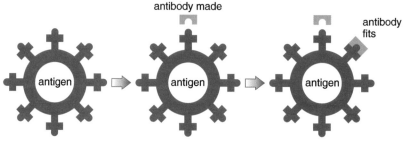
▲ *How antibodies attack antigens*

Immunity

The problem is that it can take several days for our bodies to make the antibodies. During this time we can be very ill or even die.

When we have made an antibody we recover quite quickly. We then have a copy of the antibody that can be mass-produced when needed so that if the same pathogen invades our body again, we will be immune. This is called **active immunity**.

We can avoid catching many diseases by being immunised. This involves being injected with a harmless form of the disease that has antigens.

These antigens trigger our immune system to make antibodies. If we catch the harmful form of the disease in the future, the antibodies will protect us against it.

Sometimes when we catch a disease our bodies are not capable of making the antibody in time. This is a life-threatening situation. Fortunately, doctors can sometimes inject us with antibodies that have been made by someone else. This is called **passive immunity**.

Unfortunately these antibodies from someone else do not last. In order to gain long term protection we need to produce our own antibodies to the disease.

There are always risks when foreign substances are injected into our bodies. However, these risks are always much less than the risks of catching the real disease. This is why most parents are happy for their children to be immunised against several diseases such as measles and mumps.

 d **Suggest why all children should be immunised even though there may be risks from the immunisation.**

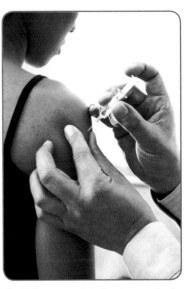

▲ *This person is being immunised*

Fighting infection

Antibiotics are drugs that are used to treat bacterial and some fungal infections. In the past, antibiotics have been used too freely. Doctors have sometimes prescribed them to patients who have a cold, knowing that antibiotics do not work against viruses. Farmers have even used them to make their animals grow more quickly. This has resulted in some bacteria becoming resistant to antibiotics. We need to use antibiotics more carefully in the future.

The superbug, MRSA, is found in a lot of UK hospitals. The bacteria is dangerous because it is resistant to nearly all known antibiotics.

Scientists are attempting to find a new antibiotic that will be effective in killing the MRSA bacteria.

keywords

active immunity • antibiotic • antibody • antigen • benign • host • malignant • parasite • passive immunity • pathogen • toxin • tumour • vector

Drug testing

One of the jobs of scientists is to discover new drugs to protect us from disease. Discovering new drugs takes a long time. It is important to ensure that they are safe to use. First they are tested on animals and then on human tissue in the laboratory.

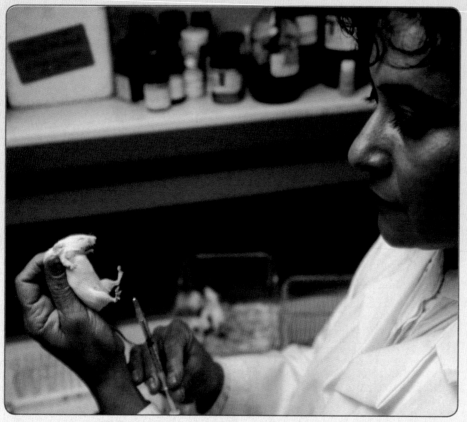

Some people object to testing drugs on animals and want to use other methods of testing such as computer modelling. Some scientists say that all drugs have side effects and it is much safer to test them on animals first.

Once the drugs are approved as safe then they are tested on humans to see how effective they are. This is done using 'blind' and 'double blind' drug trials.

Blind testing is when the patient does not know whether they are being given the real drug, or a fake pill made from substances such as flour and sugar.

This fake pill is called a placebo. The patient then has to say how effective the pill was. If those patients taking the real pill say it was effective and those patients taking the placebo say it was not effective, then the doctors know the pill is working.

Double blind testing is similar to blind testing, but this time even the doctor does not know which is the real pill and which is the placebo. Only an independent researcher knows which is which. This is so the doctor cannot give any hint to the patient about which might be the real pill.

Questions

1 Suggest why drugs are first tested on animals.

2 Why do you think some people object to testing drugs on animals?

3 Explain how blind testing of drugs works.

4 Explain why double blind testing of drugs is more reliable than blind testing.

Messages to the brain

In this item you will find out

- how the eye works and its problems

- how the brain keeps in touch with all parts of the body

- how different parts of the body communicate with each other

Imagine what it must be like to be completely blind. You would not be able to see all the colours and shapes around you. Your eyes send vast amounts of visual information to your brain.

Light enters through the eyeball where it is refracted by the cornea. The lens then focuses the light onto the retina.

The lens can change shape to focus light coming from different distances. This is called accommodation.

The lens can change shape because when the ciliary muscle contracts, it relaxes the suspensory ligaments that hold the lens, allowing the lens to become fatter. When the muscle relaxes it puts tension on the ligaments and stretches the lens to make it thinner.

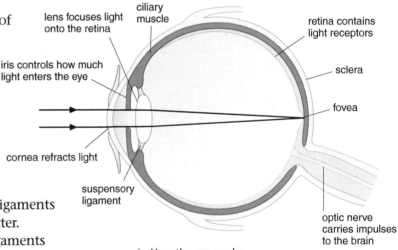
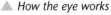

▲ How the eye works

When we get older our eye accommodation slows down and becomes poorer because the lens gets harder. This means that senior citizens often have problems when they change from looking at something close to something far away, or the other way round.

Having two eyes enables us to judge distances quite accurately – try catching a ball with one eye closed! We can do this because the brain performs some clever mathematics. As an object approaches us, the eyes have to turn inwards. The brain can use this information to judge how far away the object is. This is called **binocular vision**.

a Explain how visual information reaches your brain through your eyes.

15

Colour vision

The retina contains specialised cells that can detect red, green and blue light. These three colours enable the brain to see the world in colour. Some people inherit a condition called **colour blindness.** It is caused by a lack of the colour-detecting cells in the retina. Some people are red/green colour blind. This means that red and green look the same to them.

Problems with sight

When the eye is working properly, the cornea and the lens focus rays of light onto the retina at the back of the eye. But sometimes the eyeball or the lens is the wrong shape.

In some eyes the light is focused short of the retina. This is called **short sight**. This can be corrected by using glasses or contact lenses with a concave lens. It can also be corrected by corneal surgery using a laser. In some eyes the light is not yet in focus by the time it reaches the retina. This is called **long sight**. This can be corrected using glasses or contact lenses with a convex lens.

▼ *Concave lenses correct short sight and and convex lenses correct long sight*

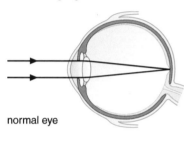

normal eye

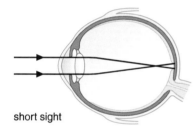

short sight

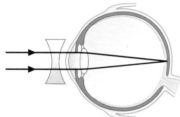

Concave lenses correct short sight.

long sight

Convex lenes correct long sight.

 b With reference to the lens, suggest why an eye may be short sighted.

The nervous system

Our nervous system consists of **sensory** nerves that carry information in the form of electrical impulses from the five senses to the brain and **motor** nerves that carry instructions to our muscles from the brain.

Each nerve consists of many nerve cells called **neurones**. Neurones are the longest cells in our body and can be over one metre in length. They act like telephone wires carrying information and instructions to and from the brain.

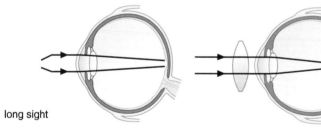

sheath axon

a sensory neurone carries instructions to our muscles

dendrites

▲ *This neurone carries information from our senses*

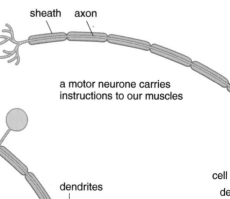

sheath axon

a motor neurone carries instructions to our muscles

cell body

dendrite

This neurone carries ▲ *instructions to our muscles*

The neurone is insulated by a **sheath**, which acts just like the plastic around the wires in an electric cable, stopping impulses travelling across from one neurone to another. The impulse travels down the **axon** from one end of the neurone to the other. The ends of each neurone have many branches called **dendrites**. This enables the neurone to connect with many other different neurones producing millions of different nerve pathways.

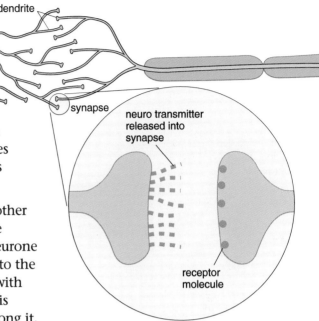

Unlike electric wires, neurones are not connected directly to other neurones. There is a gap between them called a **synapse**. The impulse crosses the synapse when it reaches the end of the neurone by triggering the release of a neuro-**transmitter** chemical into the synapse. The chemical diffuses across the synapse and binds with receptor molecules in the membrane of the next neurone. This causes the new neurone to produce an impulse that travels along it.

▲ *Synapses are the gaps between neurones*

 How is a neurone adapted to the job it does?

Reflex arc

It takes about a third of a second for an impulse from a sense organ to reach the brain and a return impulse to reach a muscle. If you have just picked up a very hot saucepan, this can be too long. Fortunately this time can be reduced by a **reflex arc**.

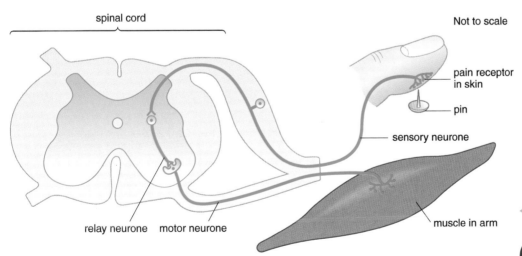

◀ *The spinal reflex arc*

In a reflex arc, the impulse goes from a **receptor**, along a sensory neurone into the spinal cord. As well as going up to the brain, a **relay** neurone connects directly to the motor neurone. The motor neurone goes to an **effector**, such as the muscle in the arm. This instructs the muscle to let go of the saucepan. By the time the brain receives the pain signal, the hand has already let go of the saucepan.

d **Suggest why the reflex arc is so important to humans.**

Seeing the optician

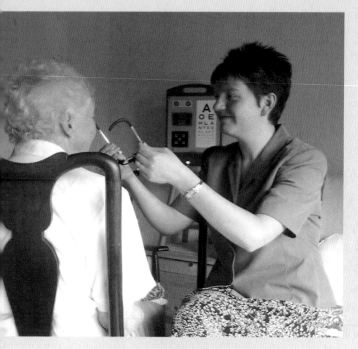

Margaret is a senior citizen. She has been having problems with her eyes so Helen the optician visits her in her home.

It is very important to have eye tests. Most people think that eye tests are performed to check whether you need to wear glasses. Eye tests, however, can tell the optician a lot about the health of the person.

The photograph below shows what Helen sees when she looks into a healthy eye.

The retina has a network of fine blood vessels.

Helen tests the pressure inside Margaret's eyeball. A blast of air is aimed at the front of the eye and the optician measures how far the cornea is pushed inwards. The higher the pressure in the eye, the less the cornea moves. If the pressure is too high, it is called glaucoma. Luckily Margaret isn't suffering from glaucoma.

Helen also checks Margaret for diabetes. Diabetes damages the blood vessels in the retina and it can lead to blindness. Unfortunately in the early stages there are no symptoms. This is why it is important that an optician spots the early signs of the disease. She can then refer the person to a doctor who will treat the patient for diabetes.

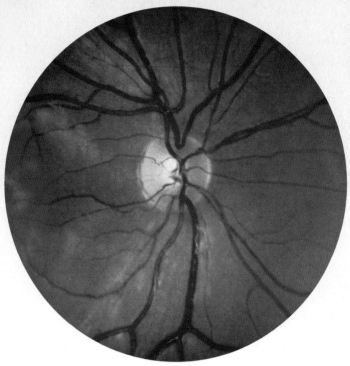

▲ Close-up of retina

Questions

1 Explain why it is important for people to have eye tests.

2 Explain why diabetes and glaucoma can damage the eye before someone realises that there is a problem.

3 Explain how Margaret is tested for glaucoma.

4 Suggest why Helen shines a bright light into Margaret's eye during an eye test.

Drugs make changes

In this item you will find out

- about different types of drugs and how they affect the body

- how drugs are classified by law

- about the effects of smoking and drinking alcohol

What do aspirin and heroin have in common? They are both painkilling **drugs**. You can buy aspirin in any chemist but if you are caught with heroin you could spend several years in prison.

So what are drugs? They are chemicals that produce changes within the body. The table shows the different types of drugs and the effects they have.

Type of drug	Effect	Example
Depressants	slows the brain down	temazepam, alcohol, solvents
Stimulants	Increases brain activity	nicotine, ecstasy, caffeine
Painkillers	reduces pain	aspirin, heroin
Performance enhancers	improves athletic performance	anabolic steroids
Hallucinogens	changes what we see and hear	LSD, cannabis

 Suggest why many people drink coffee in the morning.

Depressants and stimulants act on the synapses of the nervous system. Depressants slow down transmission across synapses. Stimulants speed up transmission across synapses.

Some drugs are social drugs. They are called this because they can be taken legally and used for recreational purposes. Alcohol and tobacco are both examples of social drugs. Even social drugs can be addictive and cause harm.

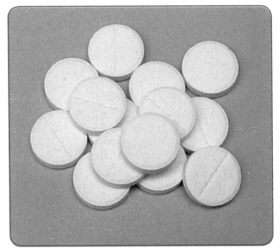

▲ *Aspirin*

▲ *Heroin*

Amazing fact

There are more synaptic pathways in the human brain than there are atoms in the universe.

Classification of drugs

The law classifies drugs into three different classes.

Class A

These are the most dangerous drugs such as heroin, cocaine, ecstasy and LSD. Illegal possession of this group carries the heaviest penalties with up to seven years in prison.

Class B

These drugs include amphetamines and barbiturates. Possession of drugs in this group can lead to up to five years in prison. Amphetamines such as 'speed' are stimulants while barbiturates are depressants.

Class C

These are mainly drugs prescribed by the doctor, and other drugs such as cannabis.

▲ A healthy lung (left), next to a smoker's lung (right)

Smoking

Smoking is highly addictive because the tobacco in cigarettes contains a drug called **nicotine**. This is why it is so hard to stop once you start.

Cigarettes also contain tar and tiny particles called particulates. The tar is carcinogenic. It damages the lungs and can cause lung cancer, while the particulates can get trapped in the lungs and block the small airways.

When cigarettes are burned carbon monoxide is produced. This is a poisonous gas that prevents the blood from carrying oxygen and can lead to heart disease.

 Look at the photograph of the healthy lung and the smoker's lung. Describe any differences you can see between them.

Effects of smoking

The trachea, bronchi and bronchioles are lined with mucus to trap dirt and microbes, and small hairs called **cilia** (ciliated epithelial cells). The job of the cilia is to waft mucus up from the lungs to the back of the throat where it is swallowed. Cigarette smoke contains chemicals that stop the cilia from working. This means that mucus accumulates in the lungs. The only way to get rid of it is to cough. This is often called a 'smoker's cough'. The build up of mucus in the bronchi can become infected and cause bronchitis.

 Describe how smoking cigarettes can result in a smoker's cough.

Look again at the photograph of the smoker's lung. You should notice that it has large holes. Smoking damages the air sacks in the lungs and produces spaces into which tissue fluid leaks. This is emphysema. It makes breathing very difficult. Chemicals in cigarette smoke can also cause cancer and heart disease. 90% of people who die from lung cancer are smokers.

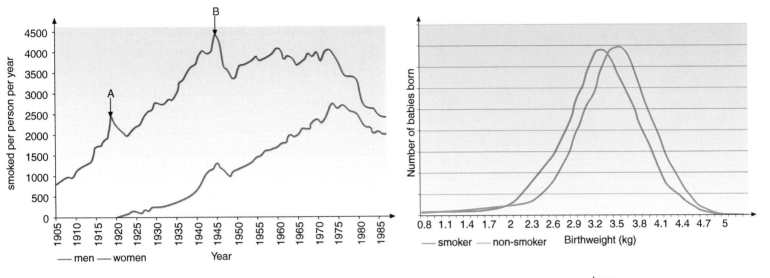

— men — women   Year

— smoker — non-smoker   Birthweight (kg)

d Look at the first graph. Suggest reasons for the two peaks, A and B.

e Look at the second graph. Explain the effect that smoking when pregnant has on the birth weight of babies.

Drinking alcohol

Alcohol is a poisonous drug that is removed from the body by the liver. Long term use can damage the liver causing a disease called **cirrhosis** of the liver. It can also damage the brain and nervous system. As the liver is responsible for a large number of the functions that take place in the body, liver damage is very serious and can lead to death. This is why long-term heavy drinking and binge drinking is so dangerous.

Alcohol consumption is measured in units. The drinks opposite all contain one or two units of alcohol each.

The recommended maximum weekly amount is 14 units for women and 21 units for men.

f If a person drank one pint of beer and two glasses of wine each day, how many units of alcohol would they drink in a week?

Alcohol also slows our reaction times. It increases the risk of having an accident when driving. This is why it is illegal to drink and drive.

beer
1 pint

2 units

wine

1 unit

spirit

1 unit

cocktail

2 units

▲ Each drink contains alcohol

Alcohol level in blood (mg/litre)	Reaction time compared with normal
0.8 the legal limit (two pints of beer)	4× slower
1.2 (3 pints of beer)	15× slower
1.6 (4 pints of beer)	30× slower

g What effect does drinking four pints of beer have on the chances of having an accident when driving a car?

keywords

cilia • cirrhosis •
depressant • drug •
hallucinogen • nicotine •
painkiller • performance
enhancer • stimulant

Should drugs be legalised?

POLICE CHIEF SAYS 'LEGALISE DRUGS'

'Legalising drugs such as cannabis would reduce crime and let us catch real criminals,' stated a Chief Constable. 'It would also reduce the amount of petty crime by drug users as they try to fund their habit.'

Arguments for:

People should be free to choose whether they take drugs or not.

Police spend too much time catching drug users.

Legalised drugs would be cheaper and safer for the users.

Cheaper drugs would mean less crime as users try to fund their habit.

Healthworkers would know who the users were and be able to help them kick the habit.

Free hypodermic needles would reduce the risk of drug users reusing them and passing on dangerous diseases such as AIDS.

Arguments against:

Increased drug use may lead to an increase in crime to pay for drugs.

People need to be protected against themselves.

Softer drugs such as cannabis may lead to using harder drugs such as heroin.

Drug use would increase if drugs were legalised.

Drugs affect reaction times. Road accidents and other types of accident would increase.

Cannabis abuse can lead to an increased risk of developing mental illnesses such as psychosis and schizophrenia.

Some people think that drugs should be legalised. Other people think that drugs should not. Some of their arguments are listed on the left.

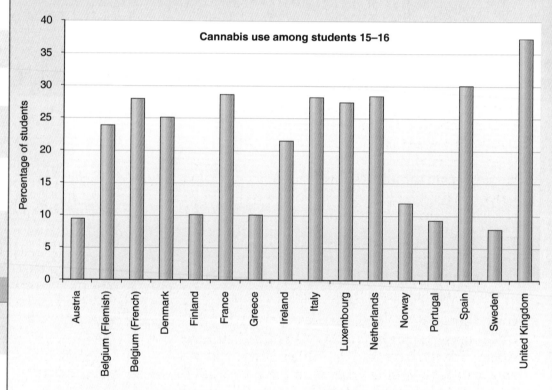

The Netherlands has a very liberal policy on drug use and many drugs are legalised. The graph shows drug use among students in different countries.

Questions

1 Some people think drugs such as cannabis should be legalised. Give reasons for and against this argument and explain your views.

2 Look at the graph. Explain what you think the effect of legalising drugs has had in the Netherlands.

3 Is it possible to draw a firm conclusion about what effect legalising drugs would have in the United Kingdom? Explain your answer.

Bodies don't like change

In this item you will find out

- why keeping a constant internal environment in our bodies is important

- how the body maintains a constant temperature

- about some hormones and what they control

In the UK, temperatures can change from below freezing to nearly 30 °C in summer. Your body has to work properly in an environment that is constantly changing.

In order for you to survive it is important that your body maintains a constant internal environment. This is called **homeostasis** and involves balancing what is taken into the body with what is given out. Your body has automatic control systems that keep temperature, water and carbon dioxide at constant levels so that cells can function at their optimum levels.

The reason why humans are so successful and have colonised every part of the planet is because of homeostasis. Most animals or plants are only found in specific areas. But humans are found everywhere.

Some animals that cannot control their own internal environment are often only found in parts of the world where conditions are just right. Others just shut down and go dormant during times when conditions are not to their liking. Even animals that can control their own internal environment, like the polar bear, are so adapted to their external environment that they can only be found in certain places in the world. All that fur would make it very hot for the polar bear if it lived at the equator. Unlike the polar bear, we can take our warm furry coats off.

▲ Clothing to protect against the sun

▲ Clothing in very cold weather

a Butterflies cannot control their temperature. It is always the same as their surroundings. Suggest why butterflies are not found at the North Pole.

b Explain homeostasis and give two examples of homeostasis in the human body.

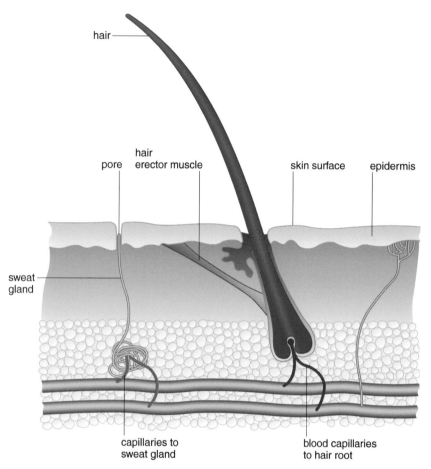

hair

pore

hair erector muscle

skin surface

epidermis

sweat gland

capillaries to sweat gland

blood capillaries to hair root

▲ *A cross-section of the skin showing the parts that regulate temperature*

Controlling body temperature

A healthy person has a constant body temperature of about 37 °C. The skin is the organ that is responsible for controlling this temperature.

This temperature control is important because the enzymes in our body work best at this temperature. Enzymes can be damaged permanently by a change in temperature. This is why a fever can be dangerous.

When we are too hot, more heat must be lost through the skin. We can do this in several ways. Blood vessels near the surface of the skin open. This is called **vasodilation**.

The skin becomes redder as warm blood is moved closer to the surface. The warm blood can then cool down as it loses heat to the surrounding air. We also start to sweat more. As the sweat evaporates, heat is removed from the skin and transferred to the environment.

When we are too cold, we must keep the heat inside our bodies. Blood vessels near the surface of the skin close. This is called **vasoconstriction**.

The skin becomes whiter as the warm blood is kept deeper inside our body. We stop sweating and start shivering. The muscles burn glucose and release heat energy into the body.

All of these changes are controlled by the brain which monitors the temperature of our blood. Because the outside temperature is changing all the time, the skin must maintain a delicate balancing act between heat lost and heat kept inside.

Amazing fact

Your skin has approximately 100 sweat glands per square centimetre.

Temperature extremes

In extreme situations the body may be unable to control its temperature. If your body temperature gets too high, you can suffer from **heat stroke**. This can lead to **dehydration** and death if it is not treated. If your body temperature gets too low you can suffer from **hypothermia** which can also lead to death.

Sex hormones

The male and female **sex hormones** are responsible for the secondary sexual characteristics that happen at puberty. Hormones are chemicals used to transmit instructions around the body.

Two of the female sex hormones are **oestrogen** and **progesterone**.

After a woman menstruates, oestrogen causes the lining of the uterus to thicken and re-grow new blood vessels.

Progesterone maintains the lining of the uterus. When the level of progesterone starts to fall towards the end of the monthly cycle, menstruation starts once more.

Oestrogen and progesterone together control other hormones produced by the pituitary gland that control the development and release of the ovum (egg) at **ovulation**.

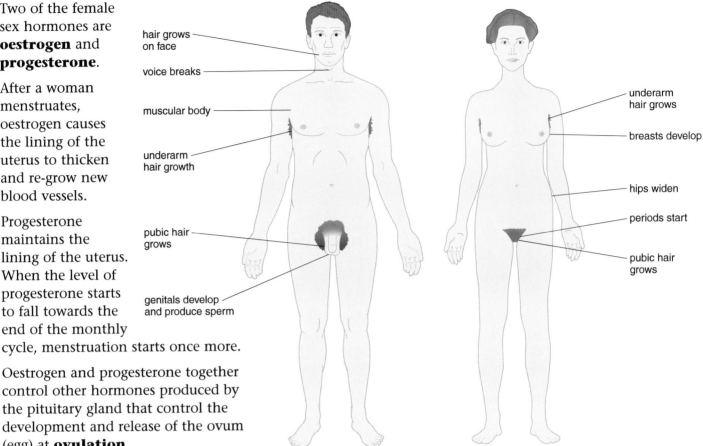

hair grows on face

voice breaks

muscular body

underarm hair growth

pubic hair grows

genitals develop and produce sperm

underarm hair grows

breasts develop

hips widen

periods start

pubic hair grows

🔺 *How sex hormones affect you at puberty*

Controlling fertility

Hormones can be used for **contraception**. Drugs such as the contraceptive pill can be used to lower **fertility**. The drugs are similar to normal female hormones and prevent the body from ovulating and releasing eggs. No eggs – no babies!

Different female hormones can be used to treat women who are infertile because they do not produce enough eggs. The extra eggs produced can be donated to other women who cannot produce any eggs of their own.

Diabetes

Some people suffer from **diabetes**. They do not produce enough of the hormone **insulin**. Insulin converts excess sugar (glucose) in the blood into glycogen that is stored in the liver.

Diabetics have to be careful that they do not eat too much sweet food. They may also need to inject themselves with the hormone insulin to help them control their blood glucose levels. The dose of insulin that they inject will depend upon their diet and how active they are.

keywords

contraception • dehydration • diabetes • fertility • heat stroke • homeostasis • hypothermia • insulin • oestrogen • ovulation • progesterone • sex hormone • vasoconstriction • vasodilation

c Suggest why a diabetic will need to inject less insulin if they have a very active lifestyle.

Negative feedback

Lucy is doing research at university into negative feedback and its applications. During homeostasis, for example, negative feedback is the control system used to maintain a constant internal environment.

When we are very hot, the brain senses the temperature of the blood passing through it. It reduces the impulses along nerves to small muscles that surround capillaries in the skin, causing them to relax. This allows more blood to flow near the surface of the skin and to lose heat.

As the blood cools, the brain senses that the temperature of the blood passing through it is dropping. It then increases the impulses to the muscles, which causes them to contract and reduce the blood flow to the surface of the skin. Blood is diverted deeper into the body and heat is conserved.

Negative feedback is a process that can also be used by industry. The diagram belows shows how it can be used in sheet metal manufacture.

Gamma ray detectors measure the thickness of the sheet of metal. If the sheet of metal is too thick then the rollers move together. If the sheet is too thin then the rollers move apart.

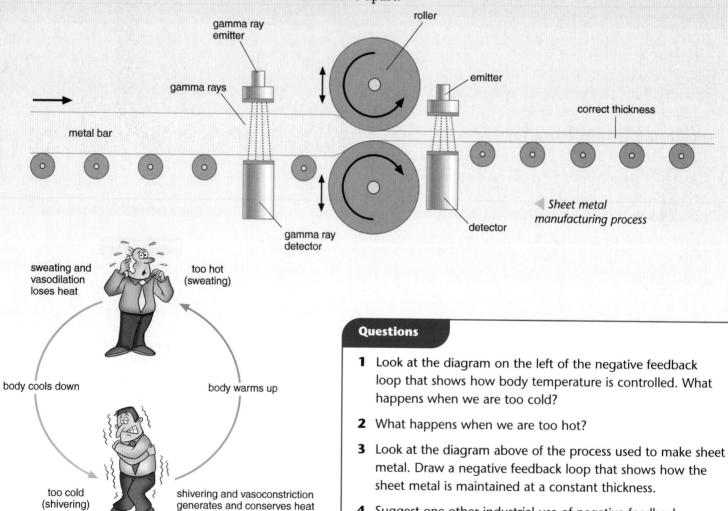

Sheet metal manufacturing process

sweating and vasodilation loses heat — too hot (sweating)

body cools down — body warms up

too cold (shivering) — shivering and vasoconstriction generates and conserves heat

Questions

1 Look at the diagram on the left of the negative feedback loop that shows how body temperature is controlled. What happens when we are too cold?

2 What happens when we are too hot?

3 Look at the diagram above of the process used to make sheet metal. Draw a negative feedback loop that shows how the sheet metal is maintained at a constant thickness.

4 Suggest one other industrial use of negative feedback.

It's all in the genes

DNA is an amazing chemical:

- it codes for all the information needed to make a new human being
- it can copy itself so that information can be passed on to future generations
- it is small enough to be stored inside the nucleus of nearly every cell in your body.

In order to fit all the DNA into each nucleus, the DNA is coiled many times to make it shorter. It is similar to a filament in a light bulb. A filament can be over 30 cm in length and coiling it up makes it shorter. The difference from DNA is that each coil is coiled again and again to make it very short indeed.

▲ DNA is very complex

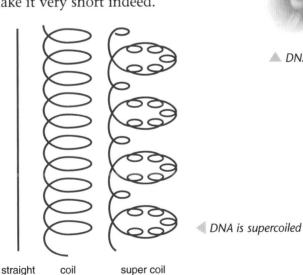

◀ DNA is supercoiled

straight   coil   super coil

Amazing fact

If the information contained by one person's DNA was written down it would fill 1000 encyclopaedias.

 Explain how such a long molecule as DNA can be squeezed into the nucleus of all our cells.

The job of DNA is to control how your cells work by controlling the production of proteins. Some of these proteins are enzymes which cause different chemical reactions to happen in your body.

▲ *Rowan has more than 30,000 genes*

Chromosomes and genes

DNA is divided into areas called **genes**. A gene codes for a single instruction. A **chromosome** is made up of long, coiled molecules of DNA. All the chromosomes contain all the information for making a new human being.

In order to code for all the information, DNA uses four different chemicals, or **bases**. These four chemicals are called A, T, C and G. Just like using letters of the alphabet to make words and sentences, it is the order of these bases that stores the code or instructions for making a new human being. Each gene contains a different sequence of bases.

DNA only has four letters in its alphabet. This means that unlike words in English that are made up of about five or six letters, genes are made up of hundreds of bases. This makes chromosomes very long indeed. This is the four-letter **genetic code** that makes the hormone insulin.

atggccctgtggatgcgcctcctgcccctgctggcgctgctggccctctggggacctgacc
cagccgcagcctttgtgaaccaacacctgtgcggctcacacctggtggaagctctctacct
agtgtgcggggaacgaggcttcttctacacacccaagacccgccgggaggcagaggacc
tgcaggtggggcaggtggagctgggcggggggccctggtgcaggcagcctgcagcccttg
gccctggagggggtccctgcagaagcgtggcattgtggaacaatgctgtaccagcatctgc
tccctctaccagctggagaactactgcaactag

Just imagine how long the sequence of bases would need to be to make baby Rowan, who has over 30,000 genes, many of which are much longer than the one for insulin.

b Explain the difference between a gene and a chromosome.

c Explain why (unlike words in English) the instructions needed to code for a gene are so long.

On or off?

Because different cells in your body use different proteins, they use different genes – the rest are switched off. For example, all nerve cells possess a complete copy of all your DNA, but they only need to use the genes that tell them how to be nerve cells. All the other genes (such as how to make haemoglobin or how to be a heart muscle cell) are not needed and are switched off.

d Explain why some genes are switched off.

Sexual reproduction

All humans have 23 pairs of chromosomes in most of their body cells (46 in total). Different species have different numbers of chromosomes, but they are always an even number.

e Suggest why the number of chromosomes is always an even number.

Human eggs and sperm (**gametes**) have half the number of chromosomes of body cells. They each have 23 single chromosomes.

When a sperm fertilises an egg during sexual reproduction the full set of 23 pairs of chromosomes is restored.

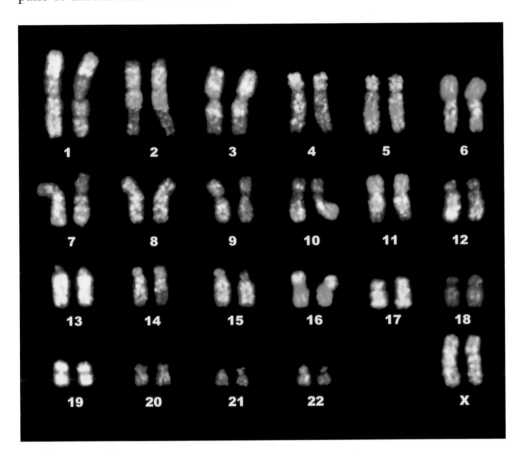

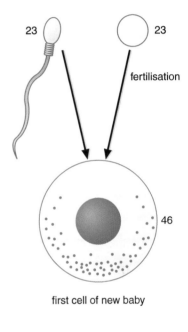

◀ *Human chromosomes*

This means that a baby contains the genetic material from both its parents.

Variety is the spice of life

The world is full of variation – apart from identical twins that have the same DNA, no two humans look the same. Some of this variation is caused by us inheriting different combinations of genes from our parents. Each gamete contains 23 chromosomes but it can be any one of the 23 from each pair. That is a lot of combinations.

More variation is caused at fertilisation itself, as any one of millions of sperm can fertilise an egg. It is hardly surprising that we are all different.

keywords

base • chromosome •
DNA • gamete • gene •
genetic code

f Describe two ways that variation is brought about during sexual reproduction.

g Explain why gametes have only 23 chromosomes when most other cells in your body have 46 (23 pairs).

Why we are all different

This CD holds all the data for making a new human being

The Human Genome Project has been one of the great triumphs of science. It has been an international research effort to sequence and map all of the genes of members of our species, *Homo sapiens*. It was completed in April 2003 and, for the first time, gave us the ability to read nature's complete genetic blueprint for building a human being.

Benefits of mapping the human genome	Disadvantages of mapping the human genome
Scientists have a better understanding of genetic diseases.	Some people think it is against God or nature to alter the genetic code.
Possible cures may be found for many diseases.	New and dangerous diseases may be produced.
We will know if we are likely to develop a disease such as heart disease and cancer in later life.	Insurance companies may refuse insurance if they know that the person is at high risk of dying early.
Most scientists want the information to be freely available to everyone in the world.	Some people want to copyright some of the genes that they have discovered.

Questions

1 When was the human genome project completed?

2 List one advantage and one disadvantage in completing the project.

3 Suggest why some drug companies may want to copyright some of the genes.

4 Suggest why insurance companies may want to know about an individual's DNA.

Uniquely you

In this item you will find out

- how DNA can be altered

- how some diseases can be inherited

- about inherited characteristics

DNA is a very delicate chemical and can easily be damaged. Changes to DNA are called **mutations**. Mutations can change the sequence of bases in DNA or even remove some of the bases.

▲ *Our DNA can be altered*

Mutations are nearly always bad news. When they happen the message in the DNA becomes disrupted. Imagine you went through this book and took out some of the letters and replaced other letters with different ones. The book would very quickly become unreadable and useless. It is the same when mutations happen to DNA – the mutation prevents the production of the protein that the genes normally code for.

▼ *The stripy colours in the rose are caused by a mutated gene*

On some occasions the mutation can be useful. By pure chance, some changes do not produce gibberish but alter the message to make a different one. For example:

'Urgent message. Send more guns and ammunition.'

'Urgent message. Send more buns and ammunition.'

This can be very useful as it can produce even more variation within a species.

Mutations can be caused by many different things or they can even occur spontaneously. All of the following can cause mutations:

- ultraviolet light in sunshine or sun beds
- chemicals in cigarette smoke
- chemicals in the environment
- background radiation in the environment.

All of these factors can damage and change the sequence of bases in our DNA. Because we have so much spare unused DNA, the chances are that we will not notice most of these mutations. However, some mutations can cause diseases such as cancer.

Alleles

We know from the previous item that a baby's cells contain two complete sets of instructions, one from each of its parents. This means that there are two different versions of each gene called **alleles**. One allele comes from the mother and the other allele from the father. Obviously, the baby cannot use both of the alleles or it would end up with two of everything.

There are two different types of alleles: **dominant** alleles and **recessive** alleles. When someone has two alleles that are the same, they are called **homozygous**. If they have two different alleles they are called **heterozygous**.

Dominant alleles are the instructions that are used. Recessive alleles are only used if someone is homozygous and has two recessive alleles.

Breeding experiment

Breeding experiments can be done using pea seeds. Let us look at the gene for height in pea plants. It has two alleles. The recessive allele is for short plants and the dominant allele is for tall plants. We will use the capital letter **T** for the dominant tall and the lower case **t** for the recessive short.

If we cross a homozygous tall plant with a homozygous short plant, all of the offspring would be Tt. Because T is dominant for height, all the seedlings would grow tall.

This type of cross using just one character such as height is called a monohybrid cross.

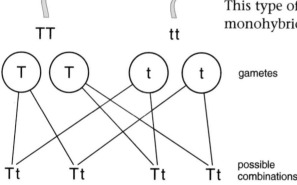

▲ Breeding experiment with pea plants

	tall	
	T	**T**
t	Tt	Tt
t	Tt	Tt

short

▶ A monohybrid cross

It is easier to see what is happening if we use a genetic diagram, such as the one above. This is called a punnet square.

Is it possible for two tall pea plants to produce seeds that will grow into short pea plants? The following genetic diagram on the left shows that it is.

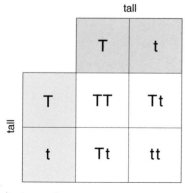

	tall	
	T	**t**
T	TT	Tt
t	Tt	tt

tall

▲ A cross between two heterozygous tall plants

a Which of the seedlings in this diagram will be tall?

b Draw a punnet square to show what will happen when a plant with alleles Tt is crossed with a plant with alleles tt.

c What proportion of the plants will be tall?

Inherited diseases

Sometimes a mutation happens in the DNA found in one of the gametes. When this happens, the mutation can be passed on to the next generation. Fortunately such mutations are usually recessive. This is because if one parent donates a faulty allele, the allele from the other parent will probably not be damaged and can provide the correct instructions.

However, problems can arise if both parents carry the same faulty gene. The punnet square on the right shows what happens when both parents who are healthy, carry the faulty gene that causes cystic fibrosis.

One in four children will have the disease and half the children will carry the recessive allele (even though they are normal and healthy).

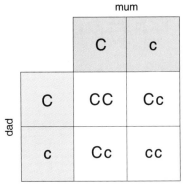

mum

	C	c
C	CC	Cc
c	Cc	cc

C = the normal healthy allele

c = the faulty allele

d Write down the combination of alleles that belong to the carriers of the disease.

Boy or girl?

Sex inheritance is controlled by a whole chromosome, rather than a single gene. In humans, males have two different chromosomes called X and Y. The X chromosome is larger than the Y chromosome.

Females have two chromosomes that are the same, called X and X.

This punnet square shows how sex is inherited.

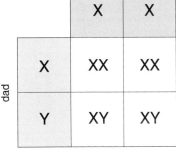

mum

	X	X
X	XX	XX
Y	XY	XY

e What is the proportion of boys born compared to girls?

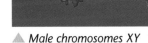

▲ Male chromosomes XY

Variation

We have seen that mutation is a major cause of variation. From the breeding experiments we can see that fertilisation of gametes also leads to variation. When gametes are produced they all contain a different combination of chromosomes. These three factors ensure that nearly all human beings are very different from one another.

Genes or environment?

Scientists are now trying to determine the relative importance of our genes as opposed to the environment in making us who we are. We know that it is a combination of both, but not how much each one contributes. If we get the right genes we may be good at sport, good at school or very healthy. But we also know that to be good at sport requires hours of training, to be good at exams requires at lot of revision and to be healthy requires eating the right foods, taking exercise and not smoking.

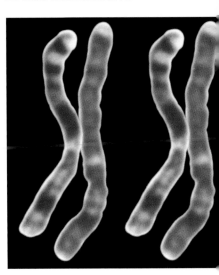

▲ Female chromosomes XX

keywords

allele • dominant • heterozygous • homozygous • mutation • recessive

33

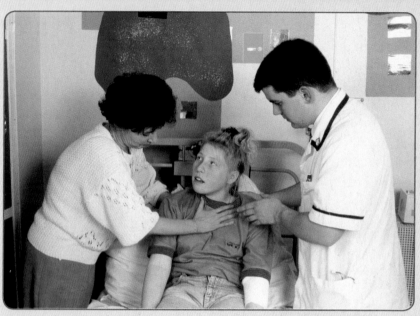

▲ This girl inherited cystic fibrosis

To know or not to know?

As we learn more and more about genetics we also learn more and more about ourselves. Soon we will understand our genes so well that we will know our risks of getting heart disease or cancer. Genetic tests are already available to see if we carry the genes for various genetic diseases.

While all this knowledge can be very useful and enable us to make informed decisions about our lives, it can also have its disadvantages.

Examiner's tip

With questions based on choice you should always give both sides of the argument in your answer.

Benefits	Disadvantages
Knowing about our genes can enable us to decide whether or not to have children. If we knew that both our partner and ourselves were carriers for the cystic fibrosis gene, we would then know that there was a one in four risk of having a child with cystic fibrosis. We could then decide whether or not to have children or take some other course of action, such as having the gametes checked, to see if they were normal.	Society has not yet decided who owns the right to know about our personal DNA.
If we knew we had a high risk of dying from heart disease we could be more careful about our lifestyle and the food that we ate.	Some insurance companies are asking clients if they have ever had any DNA tests. If they find that the person is at risk they may refuse to insure that person or raise their premiums. This will enable them to make more money for their shareholders and keep premiums down for their other clients. This means that some people may be put off having tests to find out if they are at risk from genetic disease.

Questions

1 State two advantages of knowing about our own DNA.

2 State two disadvantages of knowing about our own DNA.

3 Explain whether you think insurance companies have the right to know about the genetics of their clients.

B1a

1   *Finish the sentences by using words from the list:*

age   diastolic   mmHg   systolic

Blood pressure is measured using the units ____(1).
Blood pressure varies according to ____(2).
Blood pressure when the heart is contracting is called ____(3) pressure.   [3]

2   *Which of the following word equations best describes what happens during hard exercise?*

A   glucose + oxygen → energy
B   glucose → lactic acid + carbon dioxide + energy
C   glucose → lactic acid + energy
D   glucose → water + lactic acid + energy   [1]

3   *High blood pressure can be dangerous. Describe three serious consequences that can result from having high blood pressure.*   [3]

4   *During hard exercise, muscle fatigue can develop and the body can build up an oxygen debt.*

a   *What is this type of respiration called?*   [1]
b   *Explain the cause of muscle fatigue.*   [1]
c   *Explain why the body builds up an oxygen debt.* [1]
d   *Explain how the body repays the oxygen debt.*   [1]

5   *Fitness can be measured in different ways. Each of the following is a measure of fitness:*

strength   stamina   flexibility   agility   speed cardiovascular efficiency

Suggest and explain which type of fitness would be required for each of the following activities:

a   *marathon running*   [1]
b   *weight lifting*   [1]
c   *gymnastics*   [1]

B1b

1   *Finish the sentences by using words from the list:*
amino acids   fatty acids   glycerol   simple sugars
Carbohydrates are made up of ____(1).
Fats are made up of ____(2) and ____(3).
Proteins are made up of ____(4).   [4]

2   *Look back at the formula for calculating the recommended daily allowance of protein on page 8.*

a   *Explain how the recommended daily allowance of protein for a person is calculated.*   [1]
b   *Calculate the RDA of a person with a body mass of 80 kg.*   [1]
c   *Suggest how the RDA may vary slightly with age.*   [1]

3   *Describe how each of the following may affect a person's diet:*

a   *vegetarian*   **b**   *self-esteem*
c   *religion*   **d**   *food allergy*   [all 1]

4   *Bile plays an important roll in digestion, even though it does not contain any enzymes.*

a   *Which component of food does bile act upon?*   [1]
b   *Where is bile produced and stored?*   [1]
c   *Describe the function of bile on food.*   [1]

B1c

1   *Mosquitoes spread malaria.*

a   *What are organisms called that spread disease from one organism to another?*   [1]
b   *The following statements are in the wrong order. Write them out in the correct order.*

A   The malarial parasite develops in the mosquito.
B   The person develops malaria.
C   A harmless mosquito sucks blood from an animal with malaria.
D   A mosquito carrying the parasite sucks blood from a healthy human.   [4]

2   *Finish the sentences by using words from the list:*
active   antibiotics   antibodies   antigens damage   passive   toxins
Pathogens produce ____(1) and can cause cell ____(2).
Pathogens have ___(3) which are locked onto by ___(4).
After we recover from an infectious disease we often have ____(5) immunity to that disease.
Bacterial and fungal infections can also be treated using ____(6).   [6]

3   *MRSA is sometimes called a 'superbug'. It is resistant to almost all antibiotics. Scientists think that soon other types of bacteria will become resistant to all antibiotics.*

a   *Explain how bacteria can develop resistance to antibiotics.*   [1]
b   *Describe what can be done to prevent antibiotic resistance in bacteria.*   [1]

4   *New drugs are tested before they are used on patients.*

a   *Explain why new drugs need to be tested.*   [1]
b   *Explain why double blind trials are used in drug testing.*   [3]

B1d

1   *Explain the roll of each of the following parts of the eye:*

a   *cornea*   **b**   *iris*   **c**   *lens*
d   *retina*   **e**   *optic nerve*   [all 1]

2   Look at the diagram of the reflex arc.

List an example of each of the following from the diagram:

a   *stimulus*
b   *sensor*
c   *effector*

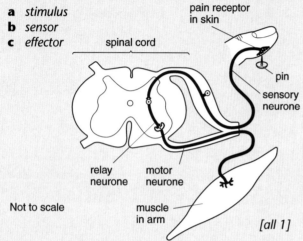

Not to scale

[all 1]

3   Neurones are highly adapted to the job that they do.

List three ways in which a neurone is adapted to its job.   [3]

4   Explain how each of the following changes its shape in order to focus light on the retina of an object approaching the eye:

a   lens   [1]
b   cilliary muscle   [2]

5   Copy out each of the following diagrams and draw a lens in front of each eye to correct the defect.   [2]

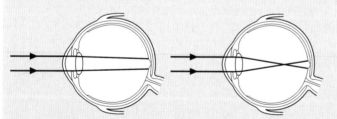

6   Synapses are the gaps between different neurones. Explain how the electrical impulse crosses the synapse from one neurone to another.   [3]

B1e

1   Finish the sentences by using word from the list:

bronchi   bronchitis   cilia   cough   mucus

Cigarette smoke stops small hairs called ____(1) from working.
The hairs normally propel sticky ____(2) up from the ____(3) to the back of the throat. A build up of sticky substance in the lungs causes ____(4) and a smokers' ____(5).   [5]

2   Alcohol consumption is measured in units.

a   How many units are there in three pints of beer? [1]
b   If a person drinks three pints of beer a night, how many units will they consume in one week?   [1]
c   Men should not consume more than 21 units each week. If a man drinks three glasses of wine and one glass of whisky each day, will he be consuming too much alcohol?   [1]
d   Women should not consume more than 14 units each week. Suggest why the unit limit for woman is less than the unit limit for men.   [1]

3   A synapse is the gap between two neurones. Explain the effect on a synapse of:

a   depressants   **b**   stimulants   [both 1]

4   Cigarette smoking causes tars and particulates to enter the lungs. Describe the effect of each of them on lung tissue.   [2]

B1f

1   The body produces a hormone that controls the body's sugar level.
a   Name this hormone.   [1]
b   What is the disease called when a person does not produce sufficient amounts of this hormone?   [1]
c   State two ways in which this disease can be treated.   [1]

2   Finish the sentences using the following words:
close   evaporates   heat   sweating   open
When the body is too hot ____(1) occurs, which ____(2) and causes the body to cool. Blood vessels near the surface of the skin ____(3) causing the skin to go red and radiate ____(4) away.   [4]

3   Oestrogen and testosterone are sometimes called the secondary sexual hormones.
a   Explain what this means.   [1]
b   Describe the effects on a young teenager of:
    i   oestrogen   **ii**   testosterone   [2]

4 The body maintains its internal environment using negative feedback.

  a Explain how body temperature is controlled by negative feedback. [1]

  b Give one example of how negative feedback can be used to control an industrial process. [1]

5 Explain the role of the following hormones on the menstrual cycle:

  a oestrogen   **b** progesterone   [both 1]

6 Explain how female hormones can be used for:

  a contraception   **b** fertility treatment   [both 1]

B1g

1 Chromosomes are found inside the nucleus.

  a How many chromosomes are found in the nucleus of most human cells? [1]

  b Is this number the same for all living organisms? [1]

  c What is unusual about the number of chromosomes found in all living organisms? [1]

2 The information required to make a human being is coded in DNA.

  a How many letters are in the DNA alphabet? [1]

  b Explain how a complete set of DNA manages to fit inside the nucleus of a cell. [1]

3 Which of the following statements about gametes is true?

  A Gametes contain the same number of chromosomes as other body cells.

  B Gametes contain twice the number of chromosomes as other body cells.

  C Gametes contain half the number of chromosomes as other body cells.

  D Gametes do not contain any chromosomes. [2]

4 Explain what is meant by the term 'one gene, one enzyme'. [2]

5 Explain why some genes in some chromosomes are 'switched off' and are not used. [1]

6 Which of the following statements ensures that sexual reproduction always produces variation in the offspring:

  A Offspring get DNA from both parents.

  B Only one female ovum is released each month.

  C Any one sperm can fertilise any one ovum.

  D Sperm are much smaller than an ovum. [2]

7 Look at the following punnet square. It shows how sex is determined.

  a Which are the sex chromosomes that determine males? [1]

  b Which are the sex chromosomes that determine females? [1]

  c Use the diagram to explain why equal numbers of boys and girls are born. [1]

	X	X
X	XX	XX
Y	XY	XY

B1h

1 Two people who could roll their tongue married and had children. Three of their children could roll their tongue and one could not.

Which condition is dominant and which is recessive? Explain your answer. [3]

2 Mutations are changes to the DNA in genes.

  a Which of the following can cause mutations to DNA: water, radiation, chemicals, sound? [2]

  b Explain whether most mutations are harmful or beneficial. [1]

3 Look at the following genetic cross between two tall pea plants.

T = Tall
t = short

	T	t
T	TT	Tt
t	Tt	tt

  a Which is dominant, tall or short? [1]

  b Which pairs of alleles are homozygous? [1]

  c What ratio of tall to short pea plants is produced by this cross? [1]

  d Write down the allele for 'tall'. [1]

4 The following cross shows the inheritance of a disease called cystic fibrosis.

  a Explain why neither the mother or the father have the condition. [1]

  b What proportion of children will have the condition? [1]

  c Is the condition dominant or recessive? [1]

	C	c
C	CC	Cc
c	Cc	cc

5 It is now possible for parents to have genetic testing carried out so that they know the odds of having a child with a genetic disorder.

Explain the advantages and disadvantages of this testing. [2]

B2 Understanding our environment

Dinosaurs are great, but why did they become extinct?

I guess they didn't look after their environment well enough and it changed so much they could not live there anymore.

- Human beings share the planet with many other species of animals and plant and yet we are now doing enormous damage to all of our environments. This damage is causing global environmental change and it is occurring at a faster rate than it has done for millions of years.

- It is only by understanding our environment that we can learn how to take more care of it. We need to understand that the environment is not ours to do with as we want. We are simply looking after it for future generations. Your children will not thank us for destroying their inheritance.

It seems strange that we are more intelligent than dinosaurs and yet we are destroying our environment but the dinosaurs didn't destroy theirs.

No, it wasn't their fault. A giant asteroid crashed into the Earth and changed their environment for them. That's why they became extinct.

Maybe we are not so bright after all.

What you need to know

- All living things are different and can be put into groups.

- Environmental, ecological and feeding relationships exist between different species of plant and animal.

- Plants are the source of all food and make it by the process of photosynthesis.

Pieces in a jigsaw

In this item you will find out

- about different ecosystems

- how data about ecosystems can be collected

- how to use keys to identify different animals and plants

We are very fortunate on Earth to have so many different types of **ecosystem**. An ecosystem is a place or habitat together with all the animals and plants that live there.

When scientists send probes to other planets, one landing site is probably much the same as any other. This makes it easier to understand and find out about the planet. Just imagine if aliens sent a probe to Earth and it landed in a desert. They might think that the whole of the planet was just one big desert.

We know more about the surface of the Moon than we know about some of the ecosystems on Earth. There are still many undiscovered **species** in places such as rainforests and the ocean depths.

Deserts, rainforests and meadows are examples of natural ecosystems. Artificial ecosystems are created by humans (examples include fields of a single crop, such as wheat or potatoes). Artificial ecosystems usually have a much lower **biodiversity**. This means there are fewer species living there.

Farmers usually have to use weed killers and pesticides to stop other species entering the ecosystem. They also usually use fertilisers to make sure the single crop species grows well before it is harvested.

▲ *Many strange species are waiting to be found at the bottom of the ocean*

▲ *Tractor spraying a field with fertiliser*

Using keys

It is important to be able to identify different animals and plants when studying an ecosystem. One way to do this would be to compare the **organism** with lots of different photographs of plants and animals to see which one it was. Unfortunately there are millions of different species, so you would need to look at millions of different photographs. This could take a long time.

a Estimate how long it would take to find a pupil at your school using a photograph, if you had to visit every class and look at every pupil.

A much better and faster way is to use a **key**. Keys work by dividing organisms into groups. Each group is then divided again and again. This may sound complicated but usually it only takes a few divisions to be able to identify an organism.

Try using this key to identify the following types of caddisfly larvae on the left.

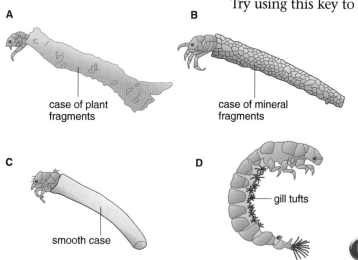

A — case of plant fragments

B — case of mineral fragments

C — smooth case

D — gill tufts — no case around abdomen

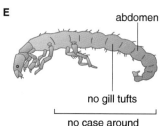

E — abdomen — no gill tufts — no case around abdomen

F — smooth case in a spiral coil

Key		
1 larvae inside a case		go to 3
larvae not inside a case		go to 2
2 has tufts on gills		Aoteapsyche (D)
does not have tufts on gills		Hydrobiosis (E)
3 has smooth case		go to 4
has rough case		go to 5
4 has straight case		Olinga (C)
has coiled case		Helicopsyche (F)
5 case made from bits of stone		Hudsonema (B)
case made from bits of plants		Triplectides (A)

b Imagine you had to identify the caddisfly larvae by looking at several million photographs. Which method, using photographs or a key, would be the quickest?

When identifying organisms by using a key, always bear in mind what you expect to find in order to check your answer. Organisms are usually found only in specific habitats. It would be most surprising if you used a key and identified a bear living on your school playing field.

Counting organisms

When you have identified all of the species in a habitat, you then need to count them. It is not usually possible to count all of the individuals in a **population**. This means you have to sample them to get an idea of how big the population is.

If you did a survey on your school playing field, one way of collecting data would be to use a quadrat. A quadrat is a metal or plastic square that encloses 0.25 square metres. The quadrat is thrown at random and all the different species found within it are identified and counted. This is much easier than doing it for the whole playing field. It is then a simple matter of multiplying the numbers counted by four to get the answer in square metres, and then multiplying the answer by the number of square metres in the playing field.

A much more accurate result can be obtained if several quadrat samples are taken and the average numbers are calculated. This is because not every quadrat will contain the same number of organisms.

▲ *Student with a quadrat*

c If a student counts eight ladybirds inside one quadrat and the playing field is 12,500 square metres, how many ladybirds would you expect to find in the playing field?

d Not every quadrat will contain eight ladybirds. Suggest how you could make the results more reliable.

e Suggest why this result will never be 100% accurate.

Quadrats cannot be used when sampling a pond or overgrown shrubland. Another method that can be used is 'capture and release'. A number of animals, such as beetles, are captured and marked with a tiny spot of paint. They are then released. Some time later, another sample of beetles is collected and examined to see how many are marked with the paint.

f Ten ladybirds are captured in a greenhouse. They are marked and then released. Sometime later, ten more ladybirds are captured and only one of them is marked. How many ladybirds do you think are living in the greenhouse?

g Suggest two different reasons why this method of counting populations may be unreliable.

◀ *A marked beetle*

keywords

biodiversity • ecosystem
• key • organism •
population • species

Then and now

Jess is doing a school project on the changing environment. In the library she has found two maps.

The map on the left shows the land around Abbey Park in Leicester in 1826. The map is over 180 years old. The map on the right is up to date.

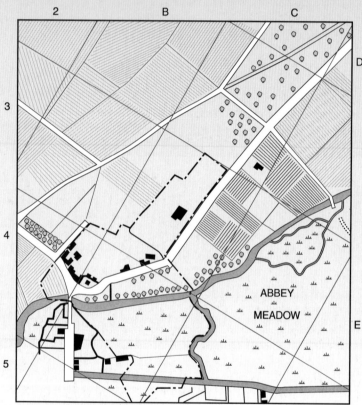

▲ The area 180 years ago

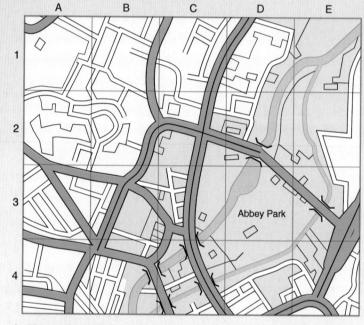

▲ The area today

Jess has also found some data on which species were around 180 years ago and which species are around today.

Questions

1. Describe two major changes that have taken place in the area.

2. What evidence suggests that the earlier map was not 100% accurate.

3. In the early map, the Abbey Meadows was a flood plain for the River Soar. List two species from the table that probably lived in the flood plain D3.

4. Suggest why the grid on the earlier map is in a different direction to the new map.

Map ref	Species	180 years ago	Today
D3	Bullrush	Lots	None
	Wheat	None	Lots
	Marsh warbler	Few	None
	Moorhen	Few	None
C4	Bluebells	Lots	Few
	Owls	Few	None
	Sparrows	Few	Lots
	Roses	None	Few
	Badgers	few	None
B3	Wheat	Lots	None
	Barley	Lots	None
	Field mouse	Lots	None
	Kestrel	None	Few

Pigeon-holing organisms

In this item you will find out

- what is meant by the word 'species'

- how living organisms are classified into different groups

- about the similarities and differences between species

▲ This frog is so new that it does not yet have a name

Scientists have just discovered a new species of frog. It is the smallest frog ever discovered in the southern hemisphere. It is only 1 cm long and can sit easily on a small coin.

A species is a group of organisms that reproduce with each other. For example, humans can only breed with other humans. This means that all humans belong to the same species called *Homo sapiens*.

Members of a species cannot usually reproduce with any other organism to produce offspring that are fertile. There are rare occasions when two organisms of different species can breed. They produce offspring called **hybrids**.

Animal hybrids are almost always infertile and cannot breed. Breeding a male donkey with a female horse produces a **mule**. The mule is sterile and is not a true species. This makes hybrids like the mule difficult to classify. It is not a donkey, nor is it a horse. It is a hybrid.

◀ The mule is a hybrid

Just like us, a species has a first and a second name. The second name is often based on the name of the person who discovered it. This way of naming an organism is called the **binomial** system of classification.

(a) Imagine that you have discovered a new species of frog. Its first name must be *Eleutherodactylus* but its second name can be anything you like.

Suggest a name for this new species.

I NAME THIS FROG FRED

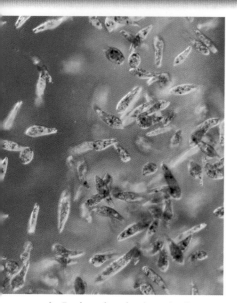

▲ *Euglena breaks the rules for animals and plants*

Plant or animal

Organisms are classified by placing them into different groups. In order to be able to do this, we need to know what criteria are used for each group. For example, animals and plants are classified into the plant or animal kingdoms using the following criteria.

It is an animal if it:	It is a plant if it:
can move independently	can only move in response to external conditions
cannot make its own food	makes its own food
is compact so that it can move about easily	is not compact and spreads out because it cannot move
does not have chloroplasts	has chloroplasts and is green

Rules and rule breakers

Unfortunately these rules do not always apply. Fungi cannot move and they are not green. This means we have to classify them in a group of their own as neither plant nor animal.

Some organisms have both plant and animal characteristics. A small microscopic pond organism called *Euglena* is green and can move. Animals can then be classified into two more groups:

- those with backbones – the **vertebrates**
- those without backbones – the **invertebrates**.

Vertebrates can then be placed into five different groups.

▲ *Worms do not have backbones and are invertebrates*

▲ *Snakes do have a backbone and are vertebrates*

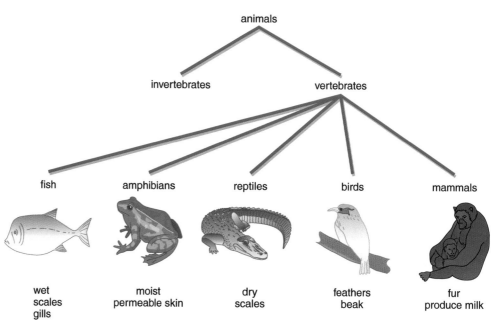

▲ *Vertebrate classification tree*

Even with vertebrates, some species break the rules. The fossil of archaeopteryx has features of both birds and reptiles.

Scientists think that birds and reptiles are both descended from archaeopteryx. All of these five different groups can be divided into smaller and smaller groups until we get to individual species.

Similar habitats

If we look at similar habits in different parts of the world, we tend to find similar species living there. On grassland in England we find grazing animals like sheep, cows and horses. In a similar habitat, like the African bush, we also find grazing animals – but this time they are zebras and gazelles. They are different animals but they eat the same food and occupy the same ecological niche as cows and sheep.

b Suggest why similar species evolve in similar habitats.

c Suggest what grazing species have evolved in the Australian outback.

▼ How apes evolved

Similar species

The apes, for example, consist of many closely-related species. Although they are all similar, they are also different from one another. The gorillas are big and heavy because they have evolved to live and gather food on the ground. Chimpanzees are smaller and lighter because they have evolved to gather food up in the trees.

Gorillas and chimpanzees have both evolved from a common ancestor in the recent evolutionary past.

Dolphins and sharks

Some species are similar to one another even though they are not closely related. This happens when two very different species are living in the same habitat. Dolphins are mammals. They breathe air using lungs and produce milk for their offspring. They are very similar to sharks, but sharks are fish. Sharks have gills and do not feed their offspring. The reason why dolphins and sharks are similar, even though they are not related, is because they both live in the same habitat. They have to swim in water so they both have to be streamlined and have fins.

▶ Dolphins and sharks have many similarities

keywords

binomial • hybrid •
invertebrate • mule •
vertebrate

From chaos into order

Eric and Sam are on a school field trip to one of the streams near their school. They are using nets to catch the insects found in the stream. Once they catch them they make notes and drawings of their features and then put the insects back. When they are back in the classroom they classify the insects.

To do this, they divide the insects into two groups. They then take one of those groups and divide into two more groups. Eric and Sam repeat the process until all the insects are classified. They know that there are many different ways of putting the insects into different groups. When we classify living organisms, we try to choose those organisms that are closely related in terms of their evolution and put them into the same group.

Questions

1 What choice could Eric and Sam make when they divide them into the first two groups?

2 How many times can they divide them into groups until they are all classified?

3 Can you think of any other way of classifying the insects. What would be your first choice this time?

4 Suggest why classifying flowers based on their colour is not as good as classifying them based on the structure of the flower.

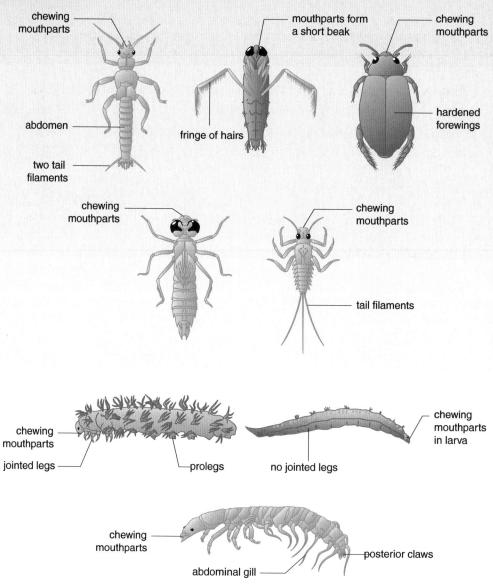

Plant magic

In this item you will find out

- about photosynthesis and how glucose can be converted into other substances

- how the rate of photosynthesis can be increased and the effect of limiting factors

- about the relationship between photosynthesis and respiration

Most students know that plants make food by **photosynthesis** from **carbon dioxide** taken from the air. It is hard to imagine that trees absorb tonnes of carbon dioxide when they photosynthesise and that millions of tonnes of carbon dioxide are put into the atmosphere when we burn fuel.

Even when we think about it, it is still difficult for us to imagine that wood (and all the vegetables and fruits that we eat) are made by plants, from carbon dioxide taken from the air and water taken from the soil.

 Which human activities put carbon dioxide back into the air?

Plants make food by converting carbon dioxide and water into **glucose** and oxygen. To do this they need energy from sunlight and a green chemical called chlorophyll.

$$\text{carbon dioxide} + \text{water} \xrightarrow[\text{chlorophyll}]{\text{light energy}} \text{glucose} + \text{oxygen}$$

$$6CO_2 + 6H_2O \xrightarrow[\text{chlorophyll}]{\text{light energy}} C_6H_{12}O_6 + 6O_2$$

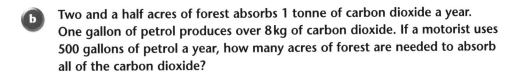 **Two and a half acres of forest absorbs 1 tonne of carbon dioxide a year. One gallon of petrol produces over 8 kg of carbon dioxide. If a motorist uses 500 gallons of petrol a year, how many acres of forest are needed to absorb all of the carbon dioxide?**

Because it is the leaves that carry out photosynthesis and absorb the carbon dioxide from the air, forests are sometimes called 'food factories'.

Amazing fact

One hectare of corn produces enough oxygen by photosynthesis for about 325 people.

Examiner's tip

The equation for photosynthesis is the same as the equation for aerobic respiration backwards.

47

Flour is mainly starch

Oil seed rape

Glucose

Glucose is very soluble and is dissolved in the plant's sap. The plant can then transport the dissolved glucose to any other part of the plant. Very often it is transported to the plant's roots for storage. Glucose is a simple sugar and is used in sweets such as chocolate bars.

Plants use the glucose for **respiration**. This releases energy for the plants to use.

Uses for glucose

Plants make much more glucose than they need for respiration. They cannot store the glucose because it is so soluble. Some plants convert the glucose into insoluble **starch**, **fats** or **oil** for storage.

When starch, fat or oil is placed inside a cell, it cannot get out because it is insoluble. Plants that store food as starch include the potato, wheat and corn. Bread flour is mainly starch.

Plants, such as oil seed rape, convert and store the glucose as oil or fat.

Glucose is also converted into **proteins** and **cellulose**. Proteins are used for growth and repair of the plant. Cellulose is the material that plant cell walls are made from.

Humans cannot digest cellulose and when we eat plants, the cellulose forms part of the roughage that passes straight through our gut.

Making photosynthesis work faster

We can make photosynthesis work faster – which is useful because it increases the rate at which farmers can grow food.

- We can increase the amount of carbon dioxide for the plants to use. We can only do this effectively in a greenhouse because carbon dioxide is a gas, Outside, the carbon dioxide would blow away.
- We can increase the amount of light reaching the plants. We can either increase the brightness of the light or the number of daylight hours.
- We can increase the temperature around the plants. Photosynthesis is a chemical reaction. Chemical reactions are faster and happen more often when the temperature increases.

c Suggest how farmers could increase the levels of carbon dioxide in a greenhouse.

Limiting factors

Carbon dioxide, light and temperature are all **limiting factors**. A lack of any one of them can limit how fast photosynthesis can go.

Look at the graphs. As light intensity increases, so does the rate of photosynthesis, but then it reaches a maximum and the graph levels out. Increasing the light further will not increase the rate of photosynthesis. But if the level of carbon dioxide is increased, then the rate of photosynthesis increases once more. The level of carbon dioxide is a limiting factor.

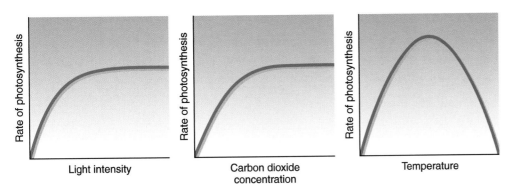

▲ Carbon dioxide is a limiting factor on the rate of photosynthesis

d What happens to the rate of photosynthesis as the amount of light increases?

e What happens to the rate of photosynthesis as the amount of carbon dioxide increases?

Respiration versus photosynthesis

Plants respire all of the time. This means that they constantly use oxygen and release carbon dioxide. During the hours of daylight they also photosynthesise.

Plants photosynthesise much faster than they respire. This means that during the day they release much more oxygen than they use and absorb much more carbon dioxide than they release. So over a 24-hour period they produce much more glucose by photosynthesis than they use for respiration.

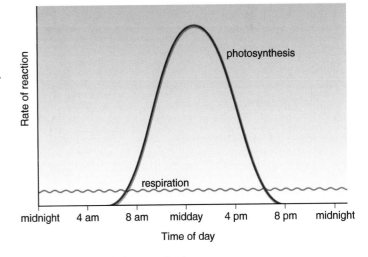

▲ Respiration and photosynthesis

f Look at the graph on the right. How many times each day is the rate of photosynthesis equal to the rate of respiration?

---

keywords

carbon dioxide •
cellulose • fat • glucose
• limiting factor • oil •
photosynthesis • protein •
respiration • starch

Increasing food production

Many commercial tomato growers use greenhouses. Modern greenhouses are very complicated places and are often controlled by a computer.

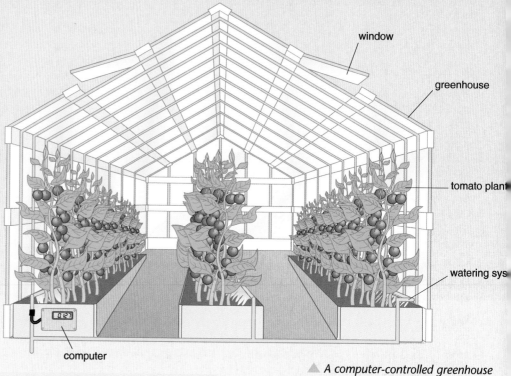

▲ *A computer-controlled greenhouse*

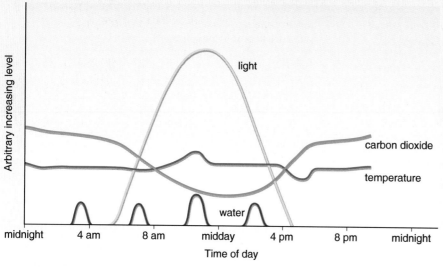

◄ *Graph of carbon dioxide, light temperature and watering in a greenhouse*

Questions

1 Suggest at what time of day the computer opened the windows.

2 State how many times during the day the computer turned on the watering system.

3 Suggest why the carbon dioxide levels fell during the hours of daylight.

4 Using only information from the graph, suggest two ways that the farmer could increase his crop of tomatoes.

5 Suggest why tomato crops grown in a greenhouse tend to be much taller than tomato crops grown outdoors.

The fight for survival

In this item you will find out

- how different animals and plants compete with each other
- about predators and prey
- about organisms that rely on other species for their survival

Each individual organism on the planet is in **competition** with all the other organisms for survival. Organisms that fail to compete successfully will die. Only successful organisms will go on to survive and breed.

Plants and animals do not just compete for food. They also compete for water, shelter, light and minerals. Because the availability of these factors varies from place to place, the distribution of organisms is also affected. For example, only organisms that can survive on small amounts of water, such as the camel, are found in the desert.

In the western world, we sometimes forget about this competition for survival because we can now control so many aspects of our own environment. Many of us have warm homes and can buy our food from shops, and now only compete for a better job or lifestyle.

Competition ensures that the population of one species does not get too big. When the population does get too big, something always happens to bring it back down to its usual size. This may be shortage of food or water, lack of space or shelter, or disease.

Sometimes when there is less competition for food, a population can explode in numbers. This happens when swarms of locusts breed and then feed on crops. Eventually the swarm runs out of food and begins to starve.

▲ In more wealthy countries humans compete for status and possessions

▲ Locusts do a lot of damage to crops

Competition between species

▲ *Red squirrel*

▲ *Grey squirrel*

Habitats can only support so many species, particularly if they are similar and competing for the same food and space. The most successful species survives and the least successful dies. This has happened to the red squirrel. When the grey squirrel was introduced to this country, it was so successful that it out-competed the native red squirrel. Red squirrels are now only found in a few isolated places in the United Kingdom. You are very lucky if you have ever seen a red squirrel in the UK.

a Suggest why the grey squirrel is more successful than the red squirrel.

b Suggest how we can ensure that populations of red squirrel continue to survive.

When two different species compete for the same part of a habitat, scientists say that they are competing for the same **ecological niche**.

Another example of an introduced species winning the battle for an ecological niche is when mink escaped from mink farms. The mink soon became wild and out-competed the native otter.

Predators and prey

Predators and **prey** both affect the size of each other's population. When the population of the predator increases, they eat more prey. This makes the prey population fall. Because there is now less food, the population of the predator falls. There are now fewer predators so the population of the prey increases again. The cycle of predator and prey populations increasing and falling is repeated over and over again. This relationship helps to maintain the balance in numbers and stops any population increase exploding out of control.

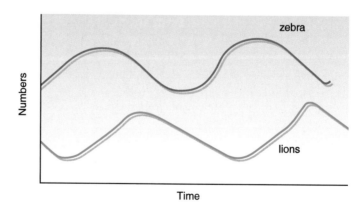

▲ *Predator and prey populations*

c Look at the graph. What do you notice about the numbers of predators compared with the numbers of prey?

d Describe one other pattern that you can see in the graph.

Parasites

Some organisms are so competitive that they can only survive by living on or in the body of other organisms. They are called parasites and the organisms they live on or in are called hosts. Examples of parasites include fleas and tapeworms. Fleas survive by living on the skin of an animal and sucking its blood. This can weaken the animal and also introduce dangerous diseases into the bloodstream.

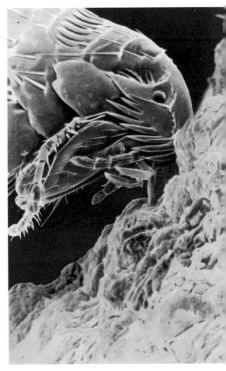

▲ Fleas feed off blood

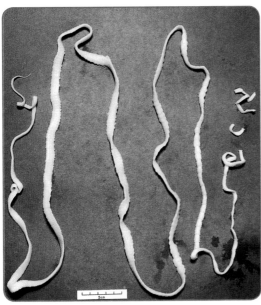

▲ Tapeworms feed off other animals' food

Tapeworms grow in the gut of an animal and feed off the food the animal eats. In severe cases they can cause a blockage in the animal's gut. They can grow up to several metres long.

Parasites always live at the expense of the host organism. They never do any good and usually harm the host. This is not always a sensible way of competing. If the host dies, the parasite will often die as well.

e **Why it is not a good idea for the parasite to kill the host on which it lives?**

▼ Oxpecker birds and giraffes live together

Mutualism

Some organisms live together. This is called **mutualism** and both organisms benefit from the relationship. One example is the oxpecker bird which eats small parasites from the fur of mammals.

Another example is nitrogen fixing bacteria that live in the root nodules of leguminous plants, such as peas, bean and clover. The plant provides the bacteria with sugars for food. The bacteria convert atmospheric nitrogen into a form that the plants can use to make protein.

This close interdependence ensures that when one organism survives and is successful so will the other one. It also determines the distribution and abundance of different organisms in different habitats.

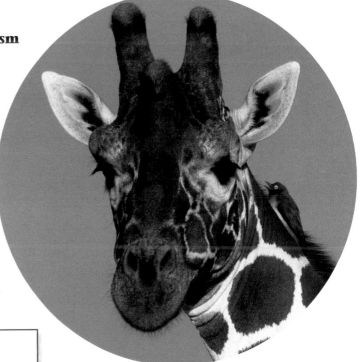

keywords

competition • ecological niche • mutualism • predator • prey

Illegal immigrants

▲ Harlequin ladybird

Britain's best-loved beetle, the ladybird, is under threat from the world's most invasive ladybird species – the harlequin ladybird.

Originally from Asia, the harlequin ladybird was first spotted in the UK in September 2004. Since then many sightings have been reported, but these have mainly been confined to the south east of Britain.

There are 46 species of ladybird in Britain and the harlequin ladybird is a potential threat to the survival of all of them. It is an extremely voracious predator that easily out-competes native ladybirds for food. When their preferred food of greenfly and scale insects is not available, the harlequin readily preys on native ladybirds and other insects such as butterfly eggs, caterpillars and lacewing larvae.

It was introduced into many countries as a biological control agent against aphid infestations in greenhouses, crops and gardens. Populations have been found in North America, France, Germany, Luxembourg, Belgium, Holland, Greece and Egypt. In France, Belgium and Holland numbers are increasing every year.

They can disperse rapidly over long distances and so have the potential to spread to all parts of the United Kingdom.

> **Examiner's tip**
>
> Don't be put off by unfamiliar examples in the examination. The principles are the same.

Key
- 2004
- 2005

▲ Distribution of harlequin ladybirds (*H. axyridis*)

Questions

1 Suggest why more sightings of the harlequin ladybird have occurred in the south east corner of the UK.

2 In what way does the harlequin ladybird out-compete the native ladybird?

3 Explain the term 'biological control'.

4 Where did the harlequin ladybird originate?

5 Suggest how the ladybird can move rapidly over long distances.

Adapt or die

In this item you will find out

- how some organisms manage to survive in harsh environments

- about different kinds of adaptations

Usually the environment changes very slowly. Over millions of years, the land we called England has been under a warm tropical ocean, had lush tropical rainforests, been covered in ice and even been a desert.

The changes take place so slowly that we do not usually notice them over the course of a lifetime. This means that animals and plants have a long time to **adapt** to the environment when it changes. Plants and animals adapt to their surroundings in order to survive. Adaptations help them compete for limited resources and increase in number.

▲ Spider on web

 a **Suggest other ways that a spider is adapted to its environment.**

Pythons are adapted so they can swallow prey that is larger than their own heads.

Spiders make sticky webs to catch prey, but the spider is adapted so that it does not stick to its own web.

When the environment changes very quickly, organisms do not have time to adapt. Some scientists think this is what may have happened to the dinosaurs.

▲ Python swallowing prey

 b **Explain why the dinosaurs became extinct when they were so perfectly adapted to their environment.**

It has been suggested that a large meteorite impacted with the Earth near Mexico. An explosion like this would have caused such rapid change to the Earth's environments that the dinosaurs would not have had time to adapt.

The more quickly that animals and plants can adapt, the more successful they will be. Their populations will have large numbers because there is less competition from other animals and plants.

How they adapt will also affect their distribution. For example, animals and plants that have adapted to live in dry environments will tend to be found in deserts.

IF DINOSAURS HAD BEEN ABLE TO ADAPT

The polar bear

layer of fat for insulation to keep heat in

thick white fur for camouflage and insulation

ears have a small surface area compared to body, this reduces heat loss

sharp claws and teeth to eat prey

strong legs for running to catch prey

large feet to spread load on snow

fur on soles of feet for insulation and grip

c Suggest why the polar bear has eyes at the front of its head rather than the sides of its head.

The camel

fat stored in a hump rather than insulating the body

bushy eyelashes and nostrils that can close to stop sand entering the eyes and air-passages

body temperature can increase so it does not sweat and lose water

large feet to spread load on sand

d The camel uses fat as a source of energy. Where does it store this fat?

e Explain why the polar bear and camel both have large feet.

The cactus

thick cuticle to reduce water loss

leaves reduced to spines to prevent water loss

green stem for photosynthesis

thick, round fleshy tissue to reduce surface area to volume ratio – to reduce water loss and store water

long roots to reach water

f Cacti also have lots of surface roots that spread out over a large area. Suggest why this is a useful adaptation in a dry desert.

Wind pollination

Insect pollination

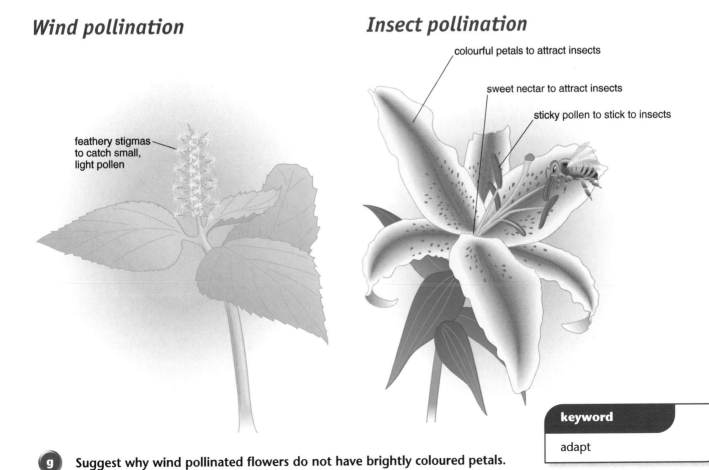

colourful petals to attract insects

sweet nectar to attract insects

sticky pollen to stick to insects

feathery stigmas to catch small, light pollen

keyword
adapt

g Suggest why wind pollinated flowers do not have brightly coloured petals.

Taking adaptation to extremes

Some organisms are so adapted to their environment that they live on the very edge of survival. One such place is deep under the ocean near 'black smokers'.

▲ *Black smoker*

Black smokers are deep sea, underwater hydrothermal vents, found in areas of undersea volcanic activity, which release large amounts of superheated solutions of minerals into the ocean. As the hot steam and water meet the cold ocean, the minerals crystallise out and form rocky deposits.

The minerals form into tall black towers that support a large variety of life such as tube-worms, giant clams and barnacles. No sunlight reaches these great depths. The organisms live in total darkness and plants cannot photosynthesise.

The energy to support this food chain comes from chemosynthetic bacteria that live on the very edge of the superheated water. They obtain their energy by chemosynthesis as they metabolise the sulfur in the hot springs. Some scientists think that this may be where life on Earth originally began.

Questions

1 Explain why green plants such as algae are not found deep in the ocean.

2 What organism provides the energy for all life found close to black smokers?

3 Explain how this organism provides energy for the food chain.

4 Suggest why it was difficult for life to have evolved in such a hostile place.

5 Suggest why scientists are so interested in life that has evolved near black smokers.

All change

- how fossils are formed and how they can be useful in understanding evolution

- what happens when environments change

- about Darwin's theory of natural selection

Humans have always asked the question 'where do we come from and how did we get here?' Over the centuries, different people have had different theories. Some people, called creationists, believe that their God created all life and that organisms were placed on Earth ready made. For example, buttercups were created as buttercups and humans created as humans.

Most scientists now think that the clues to our origin lie in the **fossil** record. However, creationists believe that even the fossil record was created by God. Scientists think that the fossil record provides evidence for the evolution of species over a period of millions of years (long before creationists believe their God created the world). The fossil that most people are familiar with and have seen for themselves is the ammonite, which scientists think existed 500 million years ago.

Sometimes when an organism dies, it becomes covered with sediment. Over many years, hard parts (such as shells and tough leaves) are gradually replaced by minerals which form the fossil.

On rare occasions, the whole organism may be preserved. This happens when bacteria that use oxygen are prevented from making the organism decay. Examples include insects embedded in amber, dinosaurs that fell into peat bogs or tar pits or mammoths that became frozen in ice.

The fossil record is not complete. This is because fossils of most organisms have not yet been discovered. Fossilisation also only very rarely occurs – most organisms do not form fossils when they die. Soft tissue usually decays and does not fossilise.

a State one piece of scientific evidence that supports evolution and one piece of scientific evidence that supports creationism.

▲ *Ammonite fossil*

▲ *Charles Darwin*

Natural selection

Organisms that are better adapted to their environments are more likely to survive. When an environment changes, many of the plant and animal species living in that environment survive or evolve, but many become extinct because they cannot change quickly enough.

In 1859, Charles **Darwin** put forward his theory of **natural selection**. Members of the same species are different from each other. This is called natural variation. These members are in competition with each other for limited resources. Because all organisms are slightly different, some are better adapted to the environment than others.

These organisms are more likely to survive and breed. This is called survival of the fittest. The offspring inherit these successful adaptations. The organisms which do not carry these characteristics die out because they are not able to compete successfully.

We now know that adaptations are controlled by genes and that these genes are inherited by the next generation.

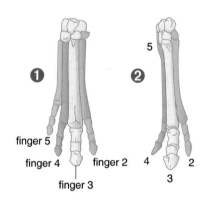

finger 5
finger 4   finger 2
finger 3

Orohippus   Miohippus

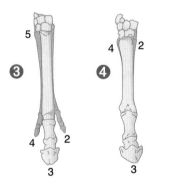

Hipparion   Equs (modern horse)

▲ *How the horse's leg has changed over time*

Evolution of the horse

In the 1870s, the palaeontologist O.C. Marsh published a description of three fossil horses that he had found in North America. Marsh thought that the horse had evolved through a series of stages from one form to another until they had evolved into the modern horse. This was called straight line evolution.

As more fossils of horses were discovered it became clear that the evolution of the horse was much more complicated than Marsh originally thought. The three fossil horses that Marsh discovered were Orohippus, Miohippus and Hipparion.

Marsh noticed that the legs of horses had evolved to allow them to run very fast.

Early horses ran on all four fingers (early horses did not have a thumb). Modern horses run on a hoof, which has evolved from just the middle finger.

b Explain what has happened to the other three fingers as the horse has evolved.

Peppered moth

Although **evolution** by natural selection usually takes place over millions of years, it is possible to see natural selection taking place over a much shorter timescale.

During the industrial revolution, heavy pollution covered the trunks of trees and bushes with black soot. Before the pollution, the grey speckled 'peppered moth' had excellent camouflage on the bark of the trees. It was very difficult for birds to spot the moths and eat them.

Can you see the peppered moth in the photograph on the right?

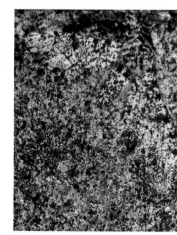

When the trees became covered in soot, the moths became much more visible and were eaten by the birds.

Fortunately, owing to natural variation, some of the moths were slightly darker in colour. These moths had better camouflage against the dark trunks and were not eaten by the birds. These moths survived, and when they bred, they passed on the genes for darker wings to their offspring. Within a few years all the moths were dark and camouflaged.

Since the clean air act, the soot pollution has disappeared and the moth has evolved back into the grey speckled variety.

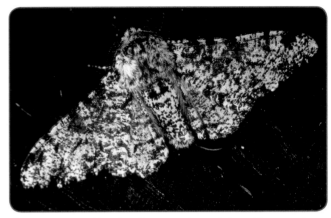
▲ *This peppered moth will not survive as it is easily spotted by predators*

 c Suggest why the moths have evolved back into the grey speckled variety.

Bugs and rats

Superbugs are resistant to nearly all of our antibiotic drugs. When antibiotics are used to treat disease, some of the bacteria may be slightly more resistant to the antibiotic than others. If these bacteria are allowed to survive, their resistance is passed on as the bacteria multiply. Soon all the bacteria are slightly resistant to the antibiotic. Because of variation, some of these bacteria have even more resistance. Within a few hundred generations, all the bacteria will be completely resistant to the antibiotic.

▲ *This peppered moth is camouflaged*

 d Suggest why doctors often prescribe two different antibiotics for very serious diseases.

Evolution and natural selection has also resulted in most rats now being resistant to the rat poison called warfarin.

Evolution of a new species

Sometimes a species may become separated into two different breeding populations. When this happens the two populations evolve independently. Sometimes they evolve so much that the differences prevent them from breeding with each other once the two populations come back together. When this happens, two new species have evolved.

 e Suggest what physical barrier could separate a species into two different breeding populations.

f Suggest what evolved differences could prevent these two populations from breeding together.

keywords

Darwin • evolution • fossil • natural selection

Different theories

In 1800, Lamarck thought that evolution occurred because characteristics that were acquired during an organism's lifetime were passed on to its offspring. If this were true, suntans and tattoos would be passed on to our children. Lamarck's theory is now discredited, but he was the first person to notice that as the environment changed so did the organisms that inhabited it. This idea enabled Charles Darwin to realise that evolution occurred because of natural selection.

The mistake that Lamarck made was to think that individual organisms controlled the characteristics that were acquired, and that these were inherited.

Like many new ideas, Darwin's theory wasn't accepted by most people. The church refused to believe that humans could be related to apes. They misunderstood Darwin and thought he said we had evolved directly from apes, rather than having a common ancestor. Scientists now think that apes and humans are the result of millions of years of evolution, and that we both evolved from a common ancestor that lived many years ago. The diagram on page 45 shows this. Even today, in the face of all the evidence, some people still believe that evolution had no part to play in the diversity of organisms that live on Earth.

It is important that when we are faced with lots of different theories we look at the evidence before we come to a conclusion. Darwin knew that the study of science was about gathering data and evidence. Some people simply accept ideas because someone else told them it is true, but scientific thought must change in the face of new evidence.

In science, some theories are disproved as more evidence is gathered that contradicts the theory. Other theories, like the theory of evolution, become stronger as the evidence gathered supports it. It is important that as we study different ideas and theories in science we have an open mind and look for evidence.

▲ *Jean Baptiste de Lamarck*

Examiner's tip

Make sure you know the difference between Darwin's and Lamarck's theories.

Questions

1 Explain why Lamarck should be credited with increasing our understanding of evolution.

2 What mistake did Lamarck make with his theory?

3 Explain why Darwin's theory of natural selection was rejected by many people.

4 Suggest why Darwin's theory about the common ancestor for apes and humans will always remain a theory and may never be proved.

Pollution problems

In this item you will find out

- what effect the human population increase is having on the environment

- how different species can be used to monitor the level of pollution

As the human population increases, more and more of the Earth's limited **resources** of fossil fuels and minerals are being used up. Once used up, they are gone forever. Some scientists estimate that within the next ten years, half of all the Earth's crude oil will have been used up. As the population grows and more resources are consumed, **pollution** also increases.

In parts of the world the human population is growing **exponentially**. This means the population in those places doubles every 53 years.

If the human population continues to grow at this rate then there will come a point when the Earth will not be able to sustain all the people. Some countries, such as China, have already taken steps to slow down population growth.

 If the human population is 6,676,871,265, calculate what the population will be after 24 hours.

Amazing fact

The human population gets bigger by more than two people every second.

◀ *The air over Los Angeles is quite polluted*

Who is doing the polluting?

Developed countries with smaller populations, consume more resources and pollute more than less developed countries with larger populations.

Place	Population (millions)	Carbon dioxide produced (billions of tonnes)
Africa	732	0.4
USA	265	4.9

 b **Suggest why the USA has fewer people, but produces more carbon dioxide than Africa.**

Pollution has serious consequences for the environment.

Global warming

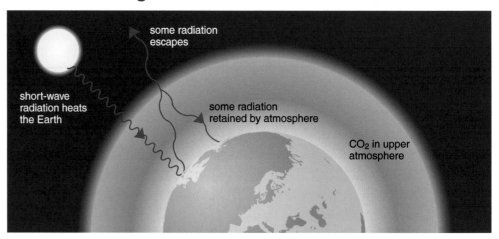

short-wave radiation heats the Earth

some radiation escapes

some radiation retained by atmosphere

CO_2 in upper atmosphere

▲ The greenhouse effect

Burning fossil fuel releases carbon dioxide into the atmosphere. We are now burning so much fuel that the level of carbon dioxide in the atmosphere is rising. Energy from the Sun hits the Earth's surface and causes warming. Normally this heat is radiated back into space. Carbon dioxide traps some of this heat energy in the atmosphere, just like the glass in a greenhouse traps heat. This causes the temperature of the Earth to rise. It is called the greenhouse effect.

Ozone depletion

The **ozone** layer is a layer of ozone gas that is found in the upper atmosphere. Ozone is normally harmful to humans, but in the upper atmosphere it absorbs most of the ultraviolet light from the Sun. Without the ozone layer, sunlight would contain so much ultraviolet light that it would not be possible for life to exist on the surface of the Earth. Unfortunately, many of the pollutants that we have produced, such as CFCs, are damaging the ozone layer. In recent years, a hole in the ozone layer has appeared over the Antarctic.

▲ The hole in the ozone layer

 c **Explain why the ozone layer is so important to life on Earth.**

Acid rain

Most fossil fuels contain small amounts of sulfur. When the fuel is burned the sulfur is released into the atmosphere as sulfur dioxide. Sulfur dioxide dissolves in rain to form **acid rain**. The acid rain kills fish (as it turns rivers and lakes into dilute acid) and it kills trees. It also reacts with limestone buildings and dissolves them away.

Measuring pollution

Pollution can be measured using biological **indicator species**. Some freshwater organisms can only survive in clean water with lots of oxygen. If they are present in the river or stream we know the water is clean and pollution free. Other organisms can survive in polluted water with very little oxygen. If they are present in the river we know that it is polluted.

Organisms found in polluted water include the blood worm and water louse. Organisms found in very polluted water include the rat tailed maggot and sludge worm.

▲ This was caused by acid rain

▲ Blood worm

▲ Rat-tailed maggot

Air pollution can be measured using lichens.

Some other lichens are much less tolerant of pollution – if you find these lichens growing you know that there is very little pollution and that the air is very clean.

keywords

acid rain • exponentially • indicator species • ozone • pollution • resource

◀ This lichen can withstand moderate levels of pollutants such as sulfur dioxide. See if you can find it on buildings around your school.

▲ *Usnea*

▲ *Parmelia caperata*

▲ *Lecanora conizaeoides*

Lichens as indicators

Different lichens can tolerate different levels of sulfur dioxide pollution. Sulfur dioxide destroys the chlorophyll that plants use for photosynthesis.

The table shows how much sulfur dioxide can be tolerated by three different lichens in $\mu g/m^3$.

Lichen	Level of sulfur dioxide tolerated ($\mu g/m^3$)
No lichens	175
Lecanora conizaeoides	125
Parmelia caperata	50
Usnea	0

Look at the map. It shows sulfur dioxide pollution in parts per billion.

The table of data and the map do not use the same units of measurement.

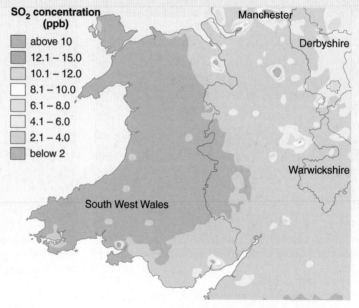

▲ *Map of Wales and Midlands showing sulfur dioxide pollution*

Questions

1 Suggest where the lichen *Usnea* may be found.

2 Suggest where the lichen *Lecanora conizaeoides* may be found.

3 The wind mostly blows form the south west. Suggest why there is less sulfur dioxide pollution in south west Wales.

4 Suggest why most lichens cannot tolerate high levels of sulfur dioxide pollution.

5 Suggest why there are high levels of sulfur dioxide in Manchester.

Extinction is forever

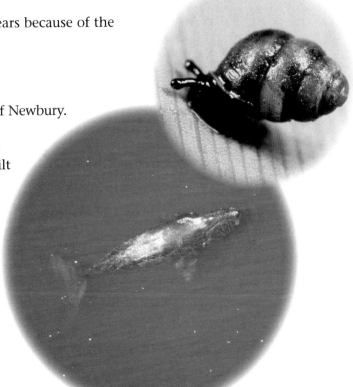

▲ Leatherback turtle

In this item you will find out

- why animals and plants become extinct
- how endangered species can be protected
- about sustainable development

When species die out, they become **extinct**. Some organisms become extinct naturally. This can happen when their environment changes owing to natural causes, such as **climate change**, or by competition with other more successful species. A species that is in danger of becoming extinct is **endangered**.

▼ Desmoulins whorl snail

 a **Describe what is meant by the phrase 'extinction is forever'.**

Some animals may become extinct in the next few years because of the effect that humans are having on the environment.

Destroying habitats

In 1995, a new bypass was planned around the town of Newbury. The habitat of an endangered rare snail needed to be destroyed to build the road. Protestors tried to stop the road from being built, but the Newbury bypass was built and opened in 1998.

Fortunately some of the snails were saved and moved to another site.

Hunting

The Northern Right Whale is on the verge of extinction. During the 1800s, whaling ships reduced the population from thousands to just a few hundred. They were called Right Whales because they were the right whales to kill by the whaling ships.

Polluting

Leatherback turtles have existed for over 100 million years. They are likely to be extinct within the next ten years. One of the greatest threats is the turtles eating ocean pollutants (such as plastic bottles) that are thrown into the sea.

▲ Northern Right Whale

▲ *Site of special scientific interest*

▲ *The red kite is a protected species*

How can humans help?

1 Protecting habitats

Habitats that contain rare or endangered species can be protected. They can be labelled sites of special scientific interest (SSSIs). This should prevent anyone developing the land and destroying the habitat.

2 Legal protection

Some species are given legal protection so that they cannot be hunted or killed. This law applies to many wild bird species.

3 Education programmes

People can be educated about how fragile and important our environment is. In this way they learn how to appreciate and respect it.

4 Captive breeding programmes

Some zoos have breeding programmes to breed rare species and release them back into the environment. Examples include birds of prey such as the red kite and the osprey.

5 Creating artificial ecosystems

Artificial ecosystems can be created for endangered species to live in. Even having your own wildlife pond can make a difference.

b Suggest another artificial ecosystem that could be created to help protect an endangered species.

Conservation programmes make sense

Conserving the environment and all the different species within it is a very sensible thing to do. The more species that exist in a habitat, the more likely the habitat is to survive. It enables the habitat to withstand small changes.

This is because there are more food chains within a complex food web and therefore more opportunities for organisms to find food.

Humans are part of the environment. When we conserve it, it means that we protect our food supply as well.

We also obtain many of our medicines from plants. There are still thousands of undiscovered plants that could potentially provide us with new life-saving drugs.

People have also become very proud of some species, such as the oak trees of England or the golden eagles in Scotland. These plants and animals form part of our cultural inheritance and we should try to conserve them.

Sustainable development

One way that the environment can be protected is by **sustainable development**. This means that whatever is removed from the environment must be replaced.

Woodland that is cut down can be sustained by replanting with young trees. The young trees grow and sustain the woodland. In the future, these trees will also be harvested and replaced.

Many woodlands now have visitor centres that are used by schools to teach about conservation, including rare and endangered species.

Fish stocks in the North Sea are being seriously over-fished. The numbers of fish, such as cod, are falling dramatically. When the fish population drops below a critical level it may not recover.

Fishermen have been given quotas to limit the number of fish that they can remove.

This should then allow the fish to breed and the fish population to recover.

Many different countries fish in the North Sea. For sustainability to work, careful planning, cooperation and agreement are required between all the different countries.

c At low tide, cockle pickers collect cockles from the sand and mud flats of Morecambe Bay. Suggest how the cockles could be conserved by sustainable development.

Sustained woodland

Examiner's tip

Make sure you know the difference between sustainable and non-sustainable development.

keywords

climate change • endangered • extinct • sustainable development

Whales in the wild

Whale watching

Whales and dolphins are mammals. They are more closely related to humans than other marine animals. There is still much that we do not understand about whales.

They can dive and survive at great depths and are able to communicate over long distances. They also migrate from one part of the world to another to follow food supplies and mate. Marine biologists are still trying to discover how whales can do all of these things.

Whales are hunted in some parts of the world. They are valued for food and the oil that they contain. Chemicals from their body are also used for manufacturing some cosmetics. Some whales are killed for research.

Because whales live far from land it is not easy to get international agreements to stop whaling. It is even harder to police the laws and ensure that some whalers do not break these international agreements.

Humans are fascinated with whales and whale watching is a valuable commercial tourist industry.

Some whales and dolphins are kept in captivity. They are used for both entertainment and research. Many are bred in captivity and are not used to swimming in the open sea.

Killer whale in captivity

Questions

1 Suggest why whale watching is such a popular tourist activity.

2 State two reasons why whales are still hunted by whaling boats.

3 State two things that we still do not fully understand about whales.

4 State two reasons why whales are kept in captivity.

5 Suggest why protecting whales is such a difficult thing to do.

6 Although whales and dolphins seem happy to perform for the public, explain whether you think they should be used in this way.

B2a

1 Describe the difference between an ecosystem and a population. [2]

2 There are many different types of ecosystem.
   a Name an ecosystem that has not yet been fully explored. [1]
   b Explain why it has not been fully explored. [1]

3 Use the key to identify the following organisms.

A   B   C   D

Key
1	has wings	go to 2
	has no wings	go to 3
2	has large antenna	butterfly
	has small antenna	housefly
3	has legs	spider
	has no legs	catterpillar [4]

4 It is almost never possible to know the exact size of a population.
   a Explain how sample size affects the accuracy of the count. [1]
   b Explain why samples may not truly represent the population. [1]

5 Explain why biodiversity in a wheat field is less than an oak woodland.

B2b

1 Describe the characteristics of the following animals:
   a fish   b amphibian   c reptile
   d bird   e mammal   [all 1]

2 Explain what the word 'species' means. [2]

3 Which of the following statements is true?

   A Similar species tend to live in similar habitats.
   B Similar species tend to live in different habitats.
   C Closely related species tend to have similar features in different habits.
   D Closely related species tend to have different features in different habitats. [2]

4 Most organisms are either animals or plants. Explain why fungi belong to neither of these two groups. [2]

5 On rare occasions, two different species can produce offspring.
   a State the word that describes an organism produced from two different species. [1]
   b State what is unusual about these types of organism. [1]
   c Explain why they are difficult to classify. [1]

6 Explain why whales and dolphins have similar features even though they are different species. [1]

B2c

1 Write down the word-equation to describe photosynthesis. [6]

2 Plants can convert glucose into many other substances. Explain how the plant uses each of the following substances:
   a glucose   b cellulose
   c protein   d starch and oils   [all 1]

3 Explain why plants respire all the time but only photosynthesise at night. [1]

4 Balance the equation for photosynthesis.

   $$\_\_\_CO_2 + \_\_\_H_2O \rightarrow C_6H_{12}O_6 + \_\_\_O_2$$ [3]

5 Look at the graph. It shows limiting factors for photosynthesis.

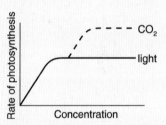

   a How many limiting factors are shown on the graph? [1]
   b Name the first limiting factor. [1]
   c Suggest how the rate of photosynthesis could be increased even further. [1]

6 The products of photosynthesis are sometimes stored.
   a Explain why glucose is soluble but storage products are not. [1]
   b State which of the following are used as storage products:

   A starch   D fat
   B simple sugars   E oils   [2]

B2d

1 Populations of organisms are not evenly distributed on the planet. List three factors that can affect their distribution. [3]

2 In a predator-prey relationship, describe what will happen when:

  a the number of predators increases   [1]
  b the numbers of prey decreases   [1]

3 Finish the sentences using the following words.

host mutualism parasites predators prey

Animals that feed on other animals are called ____(1). The animals that they feed upon are called the ____(2). Some animals live on or inside the animal that they are feeding upon. They are called ____(3). Sometimes two different species live closely together and depend upon each other. This is called ____(4)   [4]

4 Look at the following list of animals:

red squirrel hawk falcon red spotted ladybird grey squirrel yellow spotted ladybird

Put the animals in pairs to show which will be in closest competition for resources.   [3]

5 Look at the graph.

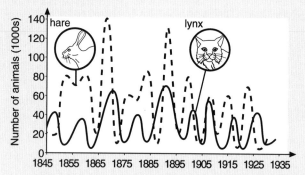

It shows the predator-prey relationship between the hare and the lynx. Describe two different patterns shown by the graph.   [2]

6 Nitrogen-fixing bacteria can be found in the roots of some plants. It is an example of mutualism. Explain why.   [2]

B2e

1 Polar bears are adapted to live in cold arctic conditions. Explain how each of the following adaptations enable the polar bear to survive:

  a white fur for   **b** layer of fat under
    camouflage   the skin
  c large feet   **d** large size   [all 1]

2 A camel is adapted to live in the desert. Explain how each of the following adaptations enable the camel to survive:

  a hump containing   **b** can allow its body
    fat   temperature to rise
  c bushy eyebrows   **d** large feet
    and hairy nostrils   [all 1]

3 The cactus is adapted to live in hot dry conditions. Explain how each of the following adaptations enable the cactus to survive:

  a large root system   **b** thick tough cuticle
  c lots of storage   **d** large volume to
    tissue   surface area ratio   [all 1]

4 Some flowers are wind pollinated.

  a Explain the difference between wind and insect pollination.   [1]
  b Explain how the following adaptations help wind pollinated flowers:
    i feathery stigmas
    ii small light pollen   [2]

5 Some flowers are insect pollinated. Explain how each of the following adaptations help insect pollinated flowers:

  a colourful petals   **b** nectar
  c sticky pollen   [all 1]

B2f

1 Fossils are the preserved remains of dead organisms.

  a Describe three different ways in which fossils can be preserved.   [3]
  b Give two reasons why the fossil record is incomplete.   [2]

2 Which of the following statements are true?

  A When the environment changes some organism may become extinct.
  B Evolution only happens when the environment changes.
  C Natural selection is when better adapted organisms die.
  D Sexual reproduction allows genes to be passed onto the next generation.   [3]

3 Put the following statements into their correct order.

  A Some moths were slightly darker and had a better chance of survival.
  B The industrial revolution produced soot that turned tree trunks black.
  C The pale moth could now be seen by predators and was eaten.
  D Pale-coloured peppered moths were camouflaged on the bark of trees.
  E Sexual reproduction produced even darker moths and after several generations all the moths were of the dark variety.   [5]

4 Describe how each of the following theories explains the fossil record:

  a Creationism   **b** Darwinism   [both 1]

5 Which of the following statements about natural selection are true?

A Natural selection can only occur if all the organisms have the same genes.

B Sexual reproduction produces variation,

C In a changing environment, organisms compete for limited resources.

D Owing to variation, some organisms are better adapted than others to the new environment.

E Species that are unable to compete may become extinct. [5]

6 Some characteristics are acquired during the life of the organism.

a Explain whether these characteristics can be inherited. [1]

b Explain the difference between Lamarck's and Darwin's theory of evolution. [1]

B2g

1 Link the following words to the correct statements:

carbon dioxide   CFCs   sulfur dioxide

A acid rain
B global warming
C destruction of ozone layer [3]

2 Look at the graph.

Describe the relationship (pattern) between population growth, use of resources and pollution.

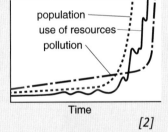

population
use of resources
pollution

Time [2]

3 The following organisms were found in three different streams:

Stream A – caddis fly and may fly larvae
Stream B – blood worm and rat tailed maggot
Stream C – stone fly larvae and water scorpion.
Which stream is the most polluted?
Explain your answer. [2]

4 Look at the following data.

Country	Population (millions)	Carbon dioxide produced (billions of tonnes)
Africa	732	0.4
America	120–170	4.9

a Explain the difference in figures between Africa and America. [1]

b Populations in parts of the world are increasing exponentially. Explain the consequences of this increase. [1]

5 Look at the graph. It shows sulfur levels in a lichen growing up to 120 km from an isolated city in America.

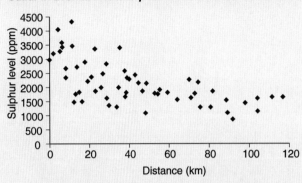

Sulfur in *Evernia mesomorpha* versus distance from source

a State the pattern between sulfur levels in the lichen and distance from the city. [1]

b Suggest why different lichens at the same distance from the city have different sulfur levels. [1]

c Suggest a possible source of the sulfur. [1]

B2h

1 Match the following reasons for extinction with the correct description:

climate change   habitat destruction   hunting pollution   competition

A fishing for North Sea cod
B two different organisms needing the same source of food
C global warming
D cutting down the rain forests
E the release of sulfur dioxide from fossil fuels

2 Humans can help endangered species. Which of the following are methods that will be successful?

A legal protection of their habitat
B increase fishing and hunting licences
C captive breeding programmes
D creating artificial ecosystems
E building new roads and towns [2]

3 Conservation is important. Give two reasons why conservation is important. Explain your answer. [4]

4 Explain why it is easier to conserve red squirrels than whales. Explain your answer. [2]

5 Describe how harvesting wood from a forest can be made into a sustainable development. [1]

C1 Carbon chemistry

- The photograph shows a scene which is all too common on our roads today and planners tell us it will get worse. Every car that is in the stationary queue of traffic is burning petrol and diesel, and emitting exhaust fumes that are damaging the environment.

- Huge amounts of crude oil are used for transport, but it is also used for other things that we take for granted – clothes, furniture and household goods. The range of materials from crude oil is vast and expanding every year.

- Crude oil is an important but finite resource. It has to be managed if supplies will be available for future generations.

The sign says I should turn my phone off at the petrol station. Why is this necessary?

Why are tests on car emissions part of the MOT test for cars?

Combustion reactions give out energy. Do all reactions give out energy?

What you need to know

- A particle model can be used to explain solids, liquids and gases, including changes of state and diffusion.

- Mixtures are composed of constituents that are not combined.

- How to separate mixtures into constituents using distillation.

Food, glorious food

In this item you will find out

- why some foods have to be cooked before they are eaten

- what cooking does to the food

- how baking powder makes a cake rise

The next time you go to a supermarket, think about all the different varieties of foods that are available. Fresh food usually has a country of origin label. If you take a look at some of them you will find that food comes from all around the world.

a Suggest how fresh food from South America gets to the UK without going bad.

Look at the food in the photograph below.

Some foods can be eaten raw or cooked, but a few foods must be cooked before we can eat them.

b Which foods in the photograph have to be cooked?

Amazing fact

The leaves of rhubarb are poisonous as they contain oxalic acid, but the stems can be eaten.

Cooking is done for a number of reasons. It:
- kills harmful microbes because it uses high temperatures
- improves the texture of food
- improves the taste and flavour of food
- makes food easier to digest.

The cooking process

Cooking food is a chemical process involving chemical reactions. We know this because the final product is different from the raw food. The cooking process is also irreversible – it is not possible to recover the raw ingredients after cooking. Finally, cooking involves an energy change.

A hen's egg is designed to accommodate a chick embryo. It is made up of three parts: the shell, the egg white and the yolk. The egg white is made up of one-eighth protein and seven-eighths water. This provides food for the growing embryo. The yolk is a yellow (or orange) oil-in-water emulsion. It is a rich source of nutrients, especially protein. It is roughly one-third fat, one-half water and one-sixth protein.

Eggs are a complete source of food for an embryo and they are a good source of protein for humans.

 Humans need proteins, fats and carbohydrates. Which food type is not in an egg?

When an egg is heated, the protein molecules in the egg change shape. The change is permanent and cannot be reversed. This is called denaturing. The photographs show an egg before and after cooking.

▲ Before cooking

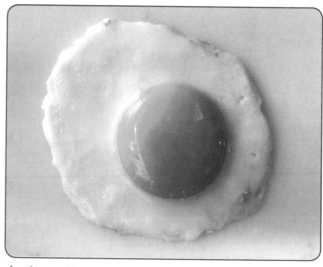

▲ After cooking

The change in the egg white is easy to see. It changes from a thick colourless liquid to a white solid, as the egg cooks. The changes to the yolk are more difficult to see, but if the egg is overcooked the yolk breaks up into a yellow powder.

Meat and potatoes

Meat is also a good source of protein. The pie diagram shows the food types in a sample of beef.

When meat is heated to 60 °C the protein molecules start to change shape. They are denatured. The meat changes colour and its texture may change depending on the method of cooking. It often shrinks.

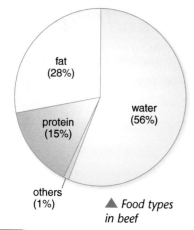

fat
(28%)

water
(56%)

protein
(15%)

others
(1%)

▲ Food types
in beef

Potatoes are **carbohydrates**. They contain cellulose, which forms the cell walls of all plants. Starch is trapped inside the potato cells. People cannot digest cellulose and uncooked starch is also difficult to digest. Cooking softens and then breaks down the cell walls releasing the starch. The starch absorbs water and becomes a gel which is more easily digested.

Baking powder

When making cakes it is important that the mixture rises during baking to give the final product a structure with lots of trapped bubbles.

When you make cakes, you use a raising agent. The simplest raising agent is **baking power** which contains sodium hydrogencarbonate. This breaks down (or decomposes) in the baking process to produce carbon dioxide.

It is the carbon dioxide which makes the cake rise and which produces lots of tiny holes.

sodium hydrogencarbonate → sodium carbonate + water + carbon dioxide

$$2NaHCO_3 \rightarrow Na_2CO_3 + H_2O + CO_2$$

▲ Trapped bubbles in a cake

A commercial baking powder contains sodium hydrogencarbonate and an acid, such as tartaric acid. When these react in the cake mixture during baking, carbon dioxide forms. Sodium tartrate also forms, but it is tasteless.

Testing for carbon dioxide

If you want to test whether a gas released during a chemical reaction is carbon dioxide, then you can bubble the gas through colourless limewater solution (calcium hydroxide). If the limewater turns cloudy, then the gas is carbon dioxide.

Limewater is calcium hydroxide solution, $Ca(OH)_2$. When carbon dioxide is bubbled through limewater, the cloudy solution is caused by calcium carbonate, $CaCO_3$.

 Write word and symbol equations for the reaction which takes place.

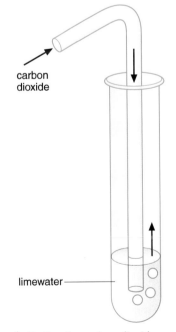

▲ Testing for carbon dioxide

Curried eggs

In 1988, Edwina Currie, the Junior Health Minister, drew attention to the problem of salmonella in British eggs. Overnight, sales of eggs fell by 60% and many egg producers went out of business. As a result, Edwina Currie had to resign.

But she had highlighted a problem and the British egg industry spent £20 billion putting it right. They used a programme of vaccination and other food safety measures.

Now British eggs are some of the safest in the world. In tests this year, only 0.34% of eggs tested were infected with salmonella. But the industry cannot relax its efforts.

Every day people in Britain eat about 30 million eggs of which about one-sixth are imported. Most of these imported eggs come from Spain and they are cheaper than British eggs. These eggs tend to be used by caterers rather than sold in supermarkets. You can tell a Spanish egg as it has ES stamped on it.

Last year 15 people died of salmonella in Britain and there were 80 known cases. Many of these cases could be traced back to Spanish eggs. Most people recover from food poisoning caused by salmonella, but certain groups are more vulnerable.

UK health authorities recommend that caterers should not use Spanish eggs in food for old people, the sick and young children. Also they should not be used in food where the egg is raw or only lightly cooked.

The Spanish authorities say that they export eggs to many other countries. They have no other complaints about salmonella in their eggs.

Questions

1 How many eggs are imported into Britain each day?

2 Suggest why caterers buy most of the imported eggs.

3 British eggs are much safer now than in 1988. Why must the authorities not relax their efforts?

4 Why are infected eggs more of a problem when they are raw or lightly cooked?

5 Which groups of people are more vulnerable?

6 Suggest why these groups are more vulnerable.

7 Why are the Spanish authorities surprised by British complaints about their eggs?

Eating by numbers

In this item you will find out

- about food additives and their advantages and disadvantages

- how 'intelligent' or 'active' packaging' is used for food preservation

- what emulsifiers do and how they do it

Food manufacturers add chemicals called **food additives** to food. There are different types of food additive. They are put in food for different reasons.

Food additive	Reason for using it
food colours	to make the food look more attractive
antioxidants	to slow down food reacting with oxygen and going bad
emulsifiers	to keep the different ingredients thoroughly mixed
flavour enhancers	to improve the flavour of the food

Food manufacturers can only use approved food additives that have been tested for safety.

How do we find out what the food we buy contains, so we know which food additives we are eating? Look at some packets of food in your kitchen. The ingredients are listed in order of their masses (with the largest first). Most additives are near the end of the list.

The label for a beef gravy flavour mix is shown below. The powder is mixed with hot water to make gravy.

 a What is the main ingredient in this beef flavour gravy mix?

Ingredients	Nutritional information	
Maize starch, Wheat flour, Flavourings, Salt, Skimmed milk powder, Flavour enhancer (E621), Colour (E150c), Onion powder, Hydrogenated palm oil, Sugar, Citric acid (E472c)	**Typical values**	**Per 100g (dry mix)**
	Energy	1371 kJ
	Protein	11.4g
	Carbohydrate	62.6g
	of which sugars	5.1g
	Fat	1.7g
	of which saturates	0.9g
	Fibre	0.6g
	Sodium	0.5g

▲ *Making gravy*

E-numbers

All approved food additives have **E-numbers**. The E-number tells us why the additive is used.

▲ *Fizzy drinks contain food colours*

E-number	Purpose	Example	Example of use
E100–199	food colours	caramel	sweets, soft drinks, jellies
E200–299	preservatives	sulfur dioxide	jams, squashes
E300–399	antioxidants	vitamin C	meat pies, salad cream
E400–499	thickeners, gelling agents, emulsifiers	starch, pectin, egg yolk	sauces, jams, salad cream and mayonnaise
E600–699	flavourings, flavour enhancers	monosodium glutamate (MSG) used in Chinese food	sweets, meat products

Antioxidants reduce the chance of oils, fats and fat-soluble vitamins from combining with oxygen and changing colour or going rancid (sour). Thickeners and gelling agents are used to thicken sauces and other foods. Emulsifiers prevent different liquids separating out.

b A fruit jelly contains E150 and E620. Suggest why it contains these two additives.

Emulsifiers

A mixture of vegetable oil and water does not mix. But it is possible to get the oil and water to mix as an **emulsion** by adding an **emulsifier** and shaking the mixture. In mayonnaise, which is an emulsion, tiny drops of oil are spread throughout the water.

The emulsifier molecules are made up of two parts. These are:

• a water-loving (**hydrophilic**) head which has an ionic charge
• an oil-loving (**hydrophobic**) tail.

The tails of the emulsifier molecules are attracted to the oil droplets. All of the heads have the same charge so there is repulsion between them. This stops the oil droplets coming together.

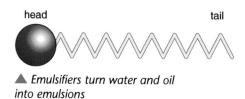

▲ *Emulsifiers turn water and oil into emulsions*

All wrapped up

When you open a bag of crisps or a packet of biscuits, do you think about the wrappings? Perhaps you don't, but there are people called packaging engineers who do. Food used to be packaged in paper or cardboard. Now it is likely to be some kind of polymer film.

The purpose of any packaging is to keep the food in good condition until it is eaten. Scientists have developed active or intelligent packaging. This means that the packaging controls or reacts to changes which are taking place inside the pack to improve the safety or quality of food. This usually means keeping out oxygen (or air) and water. Keeping out water makes it more difficult for bacteria and mould to grow. Polymer films are more effective than paper and cardboard in doing this.

Scientists have found that controlling the number of free radicals in food being stored is the key to keeping food in good condition for a long time. Free radicals are reactive atoms or groups of atoms that accelerate the breakdown of food.

There are two ways of controlling the numbers of free radicals in food.

▲ The polymer film keeps the meat fresh for longer

Oxygen scavengers

Oxygen scavengers are chemicals in the packaging that remove any oxygen in the food by reacting with it, and so preventing free radicals forming which can be harmful.

Antioxidants

Antioxidants in the packaging also combine with the free radicals in the food before they can break down the food or the packaging.

Some film packaging (for example, for fruit) needs to allow gases to leave and enter in a controlled way. The film lets oxygen in, but ethene (which causes fruit to ripen too fast) can escape.

Another example of active or intelligent packaging is found in cans that can heat or cool their contents.

> **c** What type of chemical in intelligent packaging removes oxygen from food?

Beer in cans needs to be frothy when poured out into a glass. This frothy consistency is obtained by including a small device in the can called a widget. The widget mixes air with the beer to produce the desired frothy drink.

> **d** What type of chemical in intelligent packaging removes any free radicals in the food?

> **e** Why is one type of polymer film needed for food packaging?

> **Amazing fact**
>
> Fruit and vegetables contain natural antioxidants. Pomegranates and peanuts are both very good sources of antioxidants.

> **keywords**
>
> antioxidant • emulsifier • emulsion • E-number • food additive • hydrophilic • hydrophobic

Information for the consumer

Food packaging is big business. Manufacturers spend a lot of money developing new ways to package food to keep it fresh for longer, while still being safe for us to eat.

Some people in the packaging industry believe that the way forward is in developing packaging which gives the consumer information about the food inside.

One example is packaging which changes colour depending on the temperature. This would tell the consumer if the food was getting too hot or too cold.

The VTT Technical Research Centre of Finland is developing a sensor that can be printed onto plastic packaging.

The sensor contains a substance that reacts with oxygen and lets the consumer know if food that is perishable has been packaged with oxygen.

By looking at the sensor the consumer will be able to see if the pack has been opened and is leaking oxygen.

The VTT centre is also developing methods of printing electronic data onto packs. This will help to stop counterfeit products and it will be possible to trace packs wirelessly.

It is also important in the supermarket that food is kept in the best conditions possible to prevent the food starting to breakdown. Packaging that shows when the temperature of storage is wrong by changing colour is helpful in the supermarket to show that the conditions of storage are right.

Questions

1 Suggest a use for packaging that changes colour depending on the temperature.

2 Why would a consumer want to know whether a pack is leaking oxygen?

3 Why do you think manufacturers want to be able to trace packs?

4 Suggest things that are done to ensure that a frozen chicken meal is in good condition when we buy it in the supermarket. Think about the manufacture, the transport and the storage in the supermarket.

Heaven scent

In this item you will find out

- how esters can be made and used

- why perfumes have certain properties

- why cosmetics need to be tested

An attractive smell is very powerful. The cosmetics and perfumes industries try very hard to produce smells that people will pay a lot of money to buy. Perfumes can be made by blending together natural oils (such as sandalwood or lavender), or they can be made synthetically.

But what gives perfume its smell? This is provided by a group of organic compounds called **esters**. Esters can occur naturally or they can be created synthetically.

Esters can be used as perfumes, as flavouring agents and as **solvents**. They are a very important family of organic compounds.

The mixtures of esters in a perfume are extremely expensive and the smell is very concentrated. These mixtures are diluted by adding solvents to get the perfume of the desired concentration. Perfumes can be bought in different forms, for example, eau de toilette. The more expensive products contain larger amounts of esters. Most of the perfumes produced are not used in cosmetics but are used in everyday products such as polishes or air fresheners.

Esters can even be found in sweets! The esters are added to the sweets while they are being made. The ester in pineapple chunks is methyl butanoate. This is same chemical that gives fresh pineapple its smell and taste.

▲ *Lavender oil is used in a lot of perfumes and cosmetics*

 a Describe the type of smell an ester has.

Making esters

You can make an ester by warming an organic acid with an alcohol. A drop of concentrated sulfuric acid acts as a catalyst.

organic acid + alcohol → ester + water

If you react methanoic acid and ethanol you get ethyl methanoate. A sweet raspberry smell forms.

b Finish the balanced symbol equation:

$$CH_2O_2 + C_2H_6O \rightarrow \text{?} + \text{?}$$

The table gives the names of some common esters and the acids and alcohols used to make them. It also gives the smell of the ester.

Ester	Alcohol	Acid	Smell
methyl butanoate	methanol	butanoic acid	pineapple
ethyl ethanoate	ethanol	ethanoic acid	pear drops or nail varnish remover
methyl salicylate	methanol	salicylic acid	wintergreen
methyl benzoate	methanol	benzoic acid	marzipan

c What is the name of the ester formed from ethanoic acid and methanol?

Perfumes

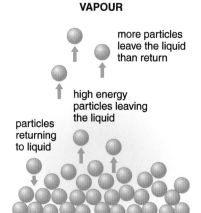

VAPOUR

more particles leave the liquid than return

high energy particles leaving the liquid

particles returning to liquid

LIQUID

Perfumes need certain properties. They need to evaporate easily, so that the perfume particles can reach the scent cells in your nose. They need to be non-toxic, so they do not poison you, and they should not irritate your skin when they are put directly on to it.

Perfumes also need to be **insoluble** in water, so they cannot be washed off easily, and they should not react with water otherwise they would react with your perspiration.

How easily a liquid evaporates is called volatility. There is only a weak attraction between the perfume particles in the liquid perfume, so it is easy to overcome this attraction. The perfume particles have lots of energy so they can escape from the other particles in the liquid. The diagram on the left shows perfume particles escaping from the liquid.

Solvents

Esters can be used as solvents. A **solution** is a mixture of a solvent and a **solute** that does not separate. Ethyl ethanoate is an ester that is a non-aqueous solvent. This means it dissolves many substances, but different substances from those dissolved by water.

Nail varnish does not dissolve in water but does dissolve in ethyl ethanoate.

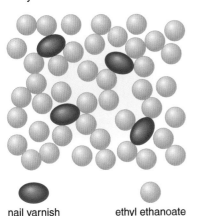

◀ *This diagram shows the particles of ethyl ethanoate and nail varnish in a solution*

nail varnish   ethyl ethanoate

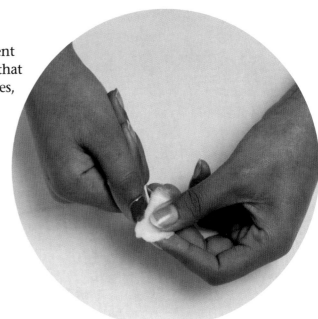

▲ *This woman is taking off nail varnish with nail varnish remover*

Water does not dissolve nail varnish. This is because the attraction between water molecules is stronger, compared with the attraction between water molecules and nail varnish particles. It is also because the attraction between nail varnish particles is stronger, compared with the attraction between water molecules and nail varnish particles.

d   **Why is it important that nail varnish does not dissolve in water?**

Testing cosmetics

Cosmetics, perfumes, other beauty products and the ingredients that are used in them, need to be tested very thoroughly before they can be sold in shops and used by people. This is so they do not harm people or cause allergic reactions.

Sometimes the products and ingredients are tested on animals. This cannot happen in the UK but it can still happen in the European Community, although it will be completely phased out by 2013. Testing on animals can only let scientists predict what effects a product or ingredient will have on humans.

However, chemicals can sometimes have different effects on different species. For example, bleach causes severe irritation to human skin, but only mild irritation to rabbit skin.

Some people feel that testing of anything on animals is wrong and should be stopped.

e   **Many people object to testing chemicals on animals. Suggest why banning testing on animals in the UK and the EU will not necessarily stop this testing.**

f   **Research shows that a chemical does not have an adverse effect on rabbits. Is this enough evidence for chemists to suggest including this chemical in cosmetics?**

keywords

ester • insoluble • solute • solution • solvent

Taking make-up off

Dr Louise Yi works in the research laboratory of a major cosmetics firm. They have just launched a new product which is a waterproof mascara.

Waterproof mascara cannot be washed off in water and the firm has been receiving complaints that their normal eye-makeup remover does not remove all traces of mascara when people take it off.

The firm wants to develop mascara removal wipes as a new product. Dr Yi is developing the cream that the wipes will contain. She is testing the effectiveness of different chemicals in dissolving the mascara.

Her results are as follows:

Solvent	Effectiveness in removing mascara (%)	Smell	Cost
Chemical A	95	unpleasant	expensive
Chemical B	76	no smell	cheap
Chemical C	87	pleasant	expensive

Questions

1 Suggest why people want to use waterproof mascara.

2 Which chemical is most effective in removing the mascara?

3 What are the drawbacks of using this chemical?

4 Which chemical do you think Dr Li should use in the new product and why?

5 Suggest what further work should be done before the wipes are manufactured and sold.

6 Some people say that cosmetics are very expensive considering that many of the ingredients are inexpensive. How can the manufacturer justify the prices they charge?

Cracking good sense

In this item you will find out

- why crude oil is a finite resource and is non-renewable

- how crude oil can be turned into useful products by fractional distillation and cracking

- about some of the environmental and political problems caused by the oil industry

▲ Crude oil is found in the North Sea but stocks are dwindling

Crude oil is a **fossil fuel** and it is a **finite resource**, which means that it will not last forever. Coal and natural gas are also fossil fuels. Fossil fuels take millions of years to make. They are **non-renewable fuels** because we are using them up faster than they can be made.

Crude oil was formed millions of years ago when small sea creatures and plants died and were buried in the seabed. The high temperatures and pressures inside the Earth turned them into crude oil. This process took place in the absence of air.

The diagram on the right shows how the crude oil is trapped with natural gas. Oil cannot escape through the rocks around the oil.

 Geologists can find where crude oil is likely to be by looking for suitable rock structures. Suggest how the oil can be extracted.

Humans have known about crude oil for thousands of years. In some places crude oil is forced onto the surface of the Earth through cracks in the crust. This then partially sets, as low boiling point hydrocarbons evaporate, and the resulting black mass is called pitch. In the time of wooden ships, such as HMS Victory, pitch used to be spread over the planks of wood.

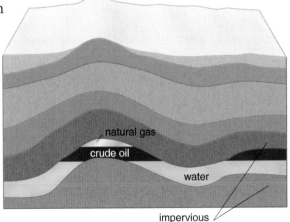

natural gas

crude oil

water

impervious rocks

 Suggest why this was done.

In Texas, 100 years ago, farmers used to burn pitch because they did not know what to do with it.

Crude oil only became valuable when people learned how to turn it into useful products.

Making crude oil useful

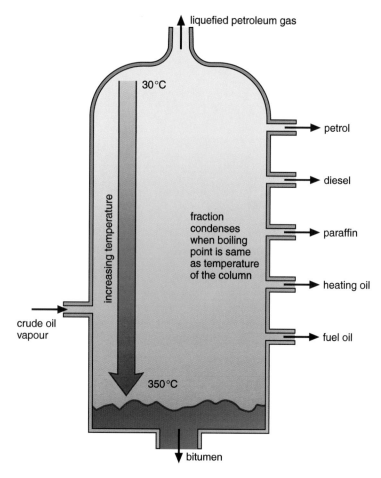

liquefied petroleum gas

30 °C

petrol

diesel

fraction condenses when boiling point is same as temperature of the column

paraffin

heating oil

increasing temperature

crude oil vapour

fuel oil

350 °C

bitumen

Crude oil is not one pure substance. It is a mixture of many different **hydrocarbons**.

Crude oil can be split into more useful products called **fractions** by **fractional distillation**. Fractions contain mixtures of hydrocarbons. The process is carried out in an oil refinery. The diagram opposite shows a fractionating column where the separation takes place.

Fractional distillation works because a fraction contains lots of substances with similar **boiling points**. Crude oil vapour enters at the bottom left of the tower. As it passes up the tower, it cools. A fraction with a high boiling point condenses and comes off at the bottom. A fraction with a low boiling point condenses and comes off at the top.

c How is crude oil vapour produced?

d Look at the diagram. Which fraction – petrol or paraffin – condenses at the lower temperature?

The **LPG** (liquefied petroleum gases) do not condense. They contain propane and butane gases.

Breaking bonds

The diagram below shows the structures of butane and octane, which are two of the fractions in crude oil. Butane contains four carbon atoms and has the molecular formula C_4H_{10}. Octane has eight carbon atoms and has the molecular formula C_8H_{18}. The carbon atoms and hydrogen atoms in hydrocarbons are bonded together with **covalent** bonds.

The covalent bonds between carbon and hydrogen are strong bonds that are not easily broken. They are stronger than the intermolecular forces between hydrocarbon molecules. When the crude oil boils during fractional distillation, the intermolecular forces between the hydrocarbon molecules are broken.

The intermolecular forces between large hydrocarbon molecules are stronger than those between smaller hydrocarbon molecules. So hydrocarbons with large molecules have higher boiling points than hydrocarbons with smaller molecules.

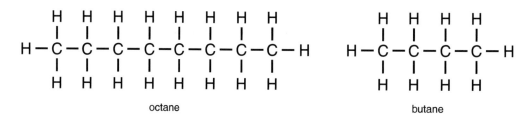

octane   butane

Supply and demand

The diagram on the right shows the uses of the different fractions in crude oil. The fractions from crude oil are used as fuels.

The table shows the supply of each fraction in crude oil and the approximate demand for each fraction by customers.

Fraction	Supply in crude oil (%)	Demand from customers (%)
LPG	2	4
Petrol	15	27
Diesel	14	21
Paraffin	14	9
Heating oil	19	14
Fuel oil and bitumen	36	25

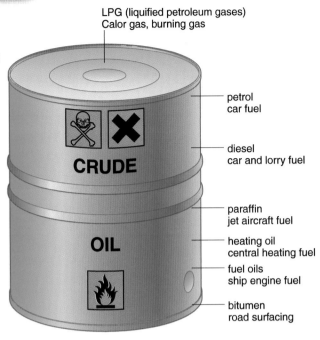

e   Look at the rows in the table for petrol and heating oil. What do you notice about the supply and the demand for these two fractions?

Cracking

Oil companies convert high boiling-point fractions (such as heating oil) into low boiling-point fractions (such as petrol) which are in greater demand. They do this by a process called **cracking**. Cracking is carried out by heating a high boiling-point fraction with a catalyst at a high temperature. There must be no air in the apparatus.

f   Why must there be no air there?

The molecules formed are smaller alkane and alkene molecules. The alkene molecules are useful because they can be used to make **polymers**.

Decane can be cracked to produce octane and ethene.

$$C_{10}H_{22} \rightarrow C_8H_{18} + C_2H_4$$

decane → octane + ethene

Environmental problems

Most of the problems caused by the oil industry come from leakage of crude oil, either at oil wells or when the oil is transported. Large sea tankers transport half of the world's crude oil. During routine tank cleaning, the oil is deliberately released into the sea. Oil can also spill into the sea when tankers unload at refineries.

Crude oil breaks down in the sea over a period of time. The lighter fractions of the oil evaporate at the water surface. Heavier fractions sink and form tar balls. These eventually break down with the help of bacteria in the water. However, this crude oil can kill sea birds and animals.

At oil wells it is usual to burn off waste gases. This also can cause atmospheric pollution.

▲ A spill from an oil tanker

keywords

boiling point • covalent • cracking • crude oil • finite resource • fossil fuel • fraction • fractional distillation • hydrocarbons • LPG • non-renewable fuel • polymer

The politics of oil

The Organisation of Petroleum Exporting Countries (OPEC) is mainly made up of Middle East Arab states and North African countries. It controls most of the world's production of oil and its price. Many of the largest deposits of oil are concentrated in countries in the Middle East where there is political unease.

The power of OPEC was shown in 1973 at the outbreak of the war between Israel and the Arab states. OPEC immediately quadrupled the price of crude oil. The economies of Western Europe and Japan, which relied on imported crude oil, went into decline. This led to recession and huge job losses.

This shows that oil-producing countries can make life very difficult for countries that rely on imported oil.

The US imports large quantities of foreign oil. This has kept costs down for consumers. A litre of petrol is much cheaper in the USA than in the UK.

This means that the USA is not in control of the amount of oil available or its price, which can lead to problems like the energy crisis of the 1970s.

During that time, foreign sources of oil refused to trade with the USA for political reasons and petrol was rationed. People with odd and even numbered license plates could buy petrol only on certain days. Americans became aware of how much they relied on foreign sources of oil. Since then, their dependence on foreign oil has increased, not decreased.

Questions

1 After 1973, oil companies were keen to extract oil from countries not in the Middle East, for example Nigeria and Venezuela. Suggest a reason for this.

2 Why did the crude oil price rise in 1973 affect Western Europe and Japan?

3 The price of crude oil does not just affect the price of fuel for cars and lorries. Suggest what other aspects of life are affected.

4 Petrol prices are much lower in the US than the UK. One reason is the large quantity of oil imported into the UK. Suggest one other reason why petrol prices are much higher in the UK.

Getting in line

In this item you will find out

- how to make polymers

- about alkanes and alkenes

- how to test for alkenes

Take a look around you. You should be able to see lots of objects that are made from polymers or plastics and you will be able to think of many others.

This photograph shows the inside of a modern car. Many of the things you can see are made of plastic. But how are polymers made?

Look at the single paper clips in the diagram. They are single units. Then look at the chain of paper clips. You can see it is made up of many paper clips joined together.

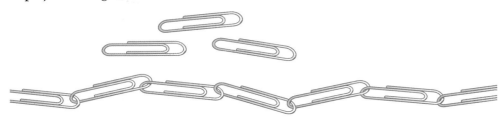

In the same way, polymer molecules are long chain molecules. In each chain there is a basic unit which repeats itself thousands of times.

The basic unit in a polymer is called a **monomer**. Many monomer molecules added together make up the polymer.

Poly(ethene) is the polymer made when lots of ethene monomer molecules are joined together. The picture on the right shows a poly(ethene) chain. You will notice that there are also some branches on the chain.

a Another polymer is poly(propene). What is the monomer for making poly(propene)?

b Styrene is a monomer. What is the polymer made from styrene?

A poly(ethene) chain ▶

Hydrocarbons

Hydrocarbons are compounds which are made up of carbon atoms and hydrogen atoms only.

c Here are four formulae. Which one is not a hydrocarbon?

CH_4 C_6H_{12} CH_2O C_2H_2

Alkanes are one family of hydrocarbons. The simplest alkanes are methane, ethane and propane.

d What do the names of the alkanes have in common?

The diagram shows the displayed and molecular formula of the first three alkanes.

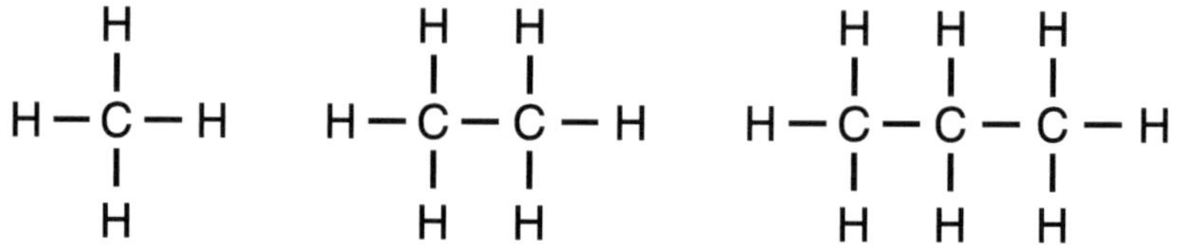

CH_4
(methane)

C_2H_6
(ethane)

C_3H_8
(propane)

All alkanes contain only single covalent bonds between their carbon atoms. Compounds that contain only single carbon-carbon bonds are called **saturated** compounds. All alkanes fit a formula C_nH_{2n+2}

Between two carbon atoms in an alkane there is a covalent bond. This is made from two electrons, one from each carbon atom. Similar covalent bonds are between carbon and hydrogen atoms. The hydrogen atoms and carbon atoms share an electron pair to form the covalent bond.

e Octane is a compound in petrol. It contains eight carbon atoms. What is the molecular formula of octane?

Alkenes are another family of hydrocarbons. The first three alkenes are ethene, propene and butene.

f What do the names of the alkenes have in common?

The diagram shows the displayed formula and the molecular formula of ethene and propene.

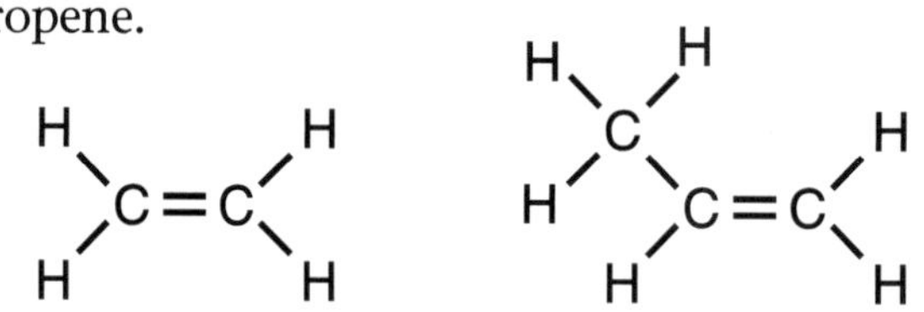

C_2H_4
(ethene)

C_3H_6
(propene)

All alkenes contain one or more double covalent bonds between their carbon atoms. Compounds that contain double covalent bonds are called **unsaturated** compounds.

g Butene has a molecular formula C_4H_8. What is the formula of an alkene containing n carbon atoms?

Testing for unsaturation

You can test whether a hydrocarbon is a saturated or an unsaturated compound. Unsaturated compounds undergo reactions that involve the breaking of one of the bonds in the carbon-carbon double bond. This is called an addition reaction, as two reactants produce a single product.

If ethene is bubbled through bromine water (a solution of bromine in water) a reaction takes place. No reaction takes place if ethane is bubbled through bromine water.

Look at the photograph on the right.

h Describe the change you see when ethene reacts with bromine.

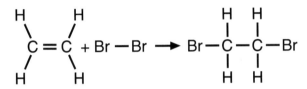

▲ *The reaction between ethene and bromine*

ethane   ethene

▲ *Testing for unsaturation*

Polymerisation

Addition polymerisation involves joining monomer molecules together by a series of addition reactions. Each monomer molecule is unsaturated, but the polymer is saturated. Addition polymerisation usually requires a high temperature and a suitable catalyst. The diagram summarises the process of polymerisation of ethene.

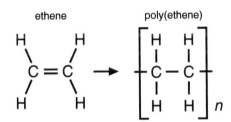

▲ *The polymerisation of ethene*

keywords

addition polymerisation
• alkane • alkene •
monomer • saturated •
unsaturated

The diagram shows ethene on the left hand side of the equation and poly(ethene) on the right. Ethene contains a double bond between the two carbon atoms and poly(ethene) contains only a single bond. Poly(ethene) does contain bonds joining the different molecules together. The letter n outside the bracket shows that a large number of ethane molecules are joined together.

Chloroethene is another unsaturated monomer that can be made into a polymer. The diagram below shows the displayed formula for chloroethene.

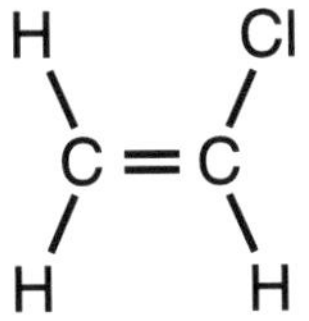

Examiner's tip

Don't forget the double bond in the monomer is lost when the polymer is formed.

i Draw the displayed formula of the polymer poly(chloroethene).

Making poly(ethene)

Poly(ethene) was first made by accident. In 1933, two chemists, Eric Fawcett and Reginald Gibson, were heating ethene gas at a very high pressure in a steel container. They did not know that some air had leaked into the apparatus. The pressure inside the apparatus kept dropping. They kept adding more ethene. When they finally opened the apparatus, they found a white waxy solid in the bottom of the apparatus.

They realised that the ethene molecules had joined together. They repeated the experiment many times over several years. They thought that oxygen was a catalyst for the joining of ethene molecules.

The poly(ethene) that they produced softened when heated and hardened again on cooling. During the Second World War, it was used as insulators for radar – which were being installed in aeroplanes. It was important that the material used was a good electrical insulator, but it also had to have a low density to reduce the weight of the plane. Without poly(ethene), the insulators would have had to be made from pottery.

Poly(ethene) was not widely available until 1950 and then it was used for making washing up bowls. Its low softening and melting temperature, however, did not make it very successful.

Chemists tried making poly(ethene) under different conditions. These samples had more desirable properties. Special metal compounds were used as catalysts. These provided a surface where the molecules could join. The resulting polymer was stronger and denser. The chains here were more regularly arranged.

Questions

1 What is the monomer used to make poly(ethene)?

2 Ethene molecules join to form poly(ethene) very slowly. Which substance in the air speeds up this process?

3 Suggest why poly(ethene) would be a better insulator for radar in aircraft than pottery.

4 Before poly(ethene), washing up bowls were usually made of painted steel. Suggest the advantages of a polymer over painted steel.

Plastics aplenty

In this item you will find out

- uses and properties of some polymers

- how the structure of a polymer relates to its properties

- problems with the disposal of polymers

Polymers are everywhere. It is now difficult to imagine life without polymers. Some of the items in the picture below are made from single polymers. Other objects are made from two or more monomers polymerised together. For example, a common form of nylon is made from two monomers polymerised together.

a Why is: (i) polystyrene used for making cups (ii) nylon used for making ropes (iii) PVC used for making wellington boots?

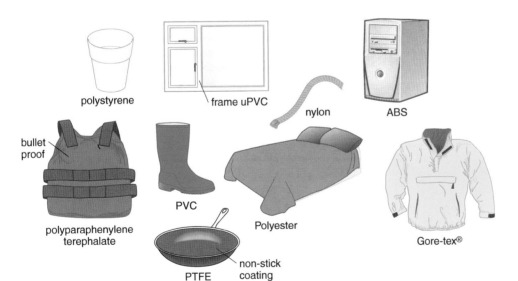

polystyrene

frame uPVC

nylon

ABS

bullet proof

PVC

polyparaphenylene terephalate

Polyester

PTFE

non-stick coating

Gore-tex®

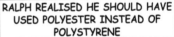

RALPH REALISED HE SHOULD HAVE USED POLYESTER INSTEAD OF POLYSTYRENE

b The computer case is made of ABS (acrylonitrile-butadiene-styrene).
(i) What three monomers are polymerised together to make ABS?
(ii) Suggest what properties ABS needs to be suitable for this use.

There is a coating of the polymer PTFE (polytetrafluoroethene) on the frying pan. It has a very slippery surface that things will not stick to.

c There are three elements in PTFE. Carbon and hydrogen are two of them. Which is the third element?

Structure of polymers

The diagram on the left shows the particles in a sample of poly(ethene).

The bonds holding the atoms together within a chain are strong covalent bonds. The intermolecular forces between the chains are weak.

Polymers that have weak **intermolecular** forces between the polymer molecules have low melting points, as little energy is needed to break the forces between the chains. They will also stretch easily because the chains can slide over each other.

This diagram shows a polymer with links between the polymer chains. This is called **cross-linking**.

The links between chains can be covalent bonds. Cross-linking changes the properties of these polymers. The chains cannot be separated by heating to a low temperature and the chains cannot slide over each other. These polymers have high melting points, are rigid and do not stretch.

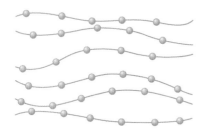

▲ Particles in poly(ethene)

▲ This polymer has links between the polymer chains

Nylon and Gore-Tex®

The anorak on page 95 is made of Gore-Tex®. Before Gore-Tex® was invented, nylon would have been used for making anoraks. Nylon is tough, lightweight and keeps water and ultraviolet light out. However, it does not let water vapour out and so sweat condenses inside the anorak.

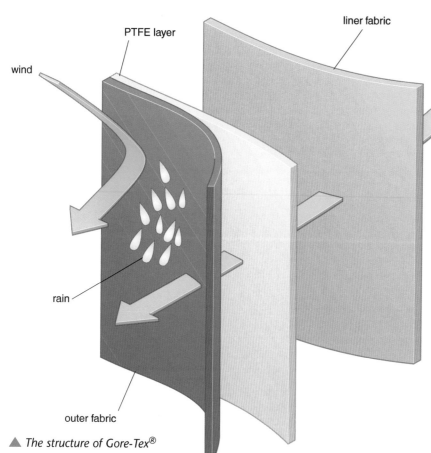

▲ The structure of Gore-Tex®

Gore-Tex® has all the positive properties of nylon but it is also breathable. Gore-Tex® is comfortable even when you get hot exercising. Any sweat is lost as water vapour through the fabric – you do not get wet even if it rains. Gore-Tex® is great for people who do outdoor activities such as cycling or mountain climbing.

Gore-Tex® is made from nylon laminated with a PTFE membrane. The holes in the PTFE are too small for water to get in, but are large enough for water vapour to pass out. The PTFE membrane is too fragile to use on its own, so it is bonded to the nylon to strengthen it.

Disposal of polymers

Polymers are widely used but there are problems. Most polymers do not decay or decompose. These polymers are called non-biodegradable. This means they can remain in landfill sites for hundreds of years. Landfill sites are filling up and it is difficult to find new ones.

Stricter and stricter legislation is being put in place to restrict the quantity of waste and the type of waste that can be put into landfill sites. Local councils are required to achieve a reduction in waste going to landfill of 25% by 2010, 50% by 2013 and 65% by 2020. This will only be achieved if more household waste, including polymers, is recycled.

Some people suggest that one solution might be to burn them, but when polymers are burned they produce toxic gases, such as dioxin. The problem could be resolved if more polymers were **recycled**. This is difficult to do because there are so many different types. Mixed polymer waste is not very useful. It can only be made into low value products, for example, insulation blocks. Sorting polymers by hand is an expensive process. The chemical industry is helping the sorting process by stamping marks on items showing what type of polymer they are made from.

The stamped label on the right indicates that the container is made from HDPE (high density poly(ethene)). If all polymer products had similar labels they could be sorted by a person picking out the different types of polymer.

▲ *The stamp on this container means it can be recycled*

d Polymers are routinely separated by hand and recycled in India but not often in the UK. Suggest why.

The future is to produce polymers that rot away quickly. Many of these are made from starch and are **biodegradable**.

keywords

biodegradable • cross-linking • intermolecular • recycled

Plastic Problem in Kenya

A report published today suggested that flimsy plastic shopping bags should be banned. Also there should be a tax on thicker plastic bags. This is to remove an increasing environmental and health problem in Kenya.

People shopping at supermarkets and shops in Nairobi alone use at least two million plastic bags each year.

The bags, many of which are so thin they are simply thrown away after one trip from the shops, have become a familiar eyesore in both urban and countryside areas. Plastic bags also block gutters and drains, choke animals and pollute the soil as they only very slowly break down.

The bags, when thrown away, can fill with rainwater. This makes a breeding ground for the malaria-carrying mosquitoes.

An estimated 4,000 tonnes of the thin plastic bags are produced each month in Kenya. About three-quarters of them are less than 30 micrometres thick. Some are as thin as seven micrometres.

A ban on bags less than 30 micrometres thick and the tax on thicker ones are among proposals aimed at reducing the use of polythene bags and providing funds for alternative, more environmentally friendly, carriers such as cotton bags.

These plans have been based on lessons learned from other countries in the world. In 2002, Ireland imposed a tax on plastic bags provided by stores and shops. It is estimated that this has reduced the use of plastic bags by 90%. In 2003, South Africa banned plastic bags thinner than 30 micrometres and introduced a plastics tax. It has seen a decrease in bag litter and a reduction in the manufacture of plastic bags.

Nairobi, 23 February 2005

Questions

1 Bags cause a litter problem when they are thrown away. Write down three other problems caused by these bags when they are thrown away.

2 Which countries already have already taken action against the use of thin plastic bags?

3 Kenya has no deposits of crude oil and has to buy all plastics abroad. Suggest reasons why cotton bags are better than plastic ones.

4 How many tonnes of plastic bags produced in Kenya each month of 30 micrometres or more can still be made?

5 Some supermarkets are experimenting with plastic bags made from a starch-based polymer. Suggest why this might be an alternative to a tax or a ban on plastic bags.

Up in flames

▼ This steam engine runs on coal

In this item you will find out

- what factors are considered when a fuel is chosen

- about the products from the combustion of carbon and carbon fuels

- about complete and incomplete combustion

Fossil fuels, such as coal, oil and gas, are used to provide most of our energy needs. Fuels are substances that react with oxygen to produce useful energy.

The three photographs show coal, oil and gas being used to power three different means of transport.

a Look at the photographs. Suggest one advantage of burning gas in the bus, rather than coal in the steam engine and diesel in the lorry.

With a choice of carbon-based fuels it is important to choose the one with the most suitable properties. We all have to make these choices in our lives. What fuel should we use to heat our houses? Should we use cars powered with petrol, diesel or LPG?

Large companies also have to make these decisions about the types of fuel that they use to produce electricity.

▲ This lorry uses diesel as its fuel

Humans are burning increasing amounts of fossil fuels. There are more and more cars on the roads and some large cars use a lot of petrol or diesel.

We are also transporting more goods to more places. Fresh fruit and vegetables are routinely flown all over the world and this form of transport also uses fuel.

▲ Buses can run on LPG gas

Choosing the best fuel

When choosing the best fuel, there are seven things that need to be considered. You can remember them if you remember the word TEACUPS:

T **toxicity** – how poisonous the fuel is
E energy value – how much energy the fuel gives out
A availability – how easy the fuel is to get hold of
C cost – how expensive the fuel is
U usability – how easy the fuel is to use
P **pollution** – how much pollution is created by the fuel
S storage – how easy the fuel is to store.

The table gives some data on three fossil fuels.

Fuel	Relative cost	Energy value	Availability	Storage	Toxity	Pollution	Ease of use
coal	cheap	medium	available	easy to store	non-toxic	smoky	difficult to catch alight
oil	expensive	high	widely available	has to be stored in a tank	leaking gas can be poisonous	little pollution	easy to burn
gas from gas main	moderate	high	only in places where there is a gas main	supply directly to house	gas non-toxic; leaks can cause explosion	little pollution	easy to burn

b Emma lives in a house in an area of a town where there is a smokeless zone. There is a gas main. Which fuel would you recommend?

c Her friend Jo lives in a block of flats. Jo is not allowed to use mains gas or bottled gas. Suggest why this is.

Burning fuels

Burning or **combustion** of all fuels needs oxygen. The combustion releases heat energy, which is very useful. When you burn a hydrocarbon fuel with lots of oxygen (air) you get carbon dioxide and water. This is called **complete combustion**.

The diagram on the opposite page shows an experiment to test what products are produced when you burn methane with lots of oxygen.

The pump sucks air through the apparatus. As the methane burns, a colourless liquid condenses and collects in the first test tube. This liquid boils at 100°C so it is water. In the second test tube a gas is bubbled through limewater. The limewater turns cloudy, so the gas is carbon dioxide. This can be shown by the following word equation:

methane + oxygen → carbon dioxide + water

d Methane has the formula CH_4. Write the balanced symbol equation for the complete combustion of methane.

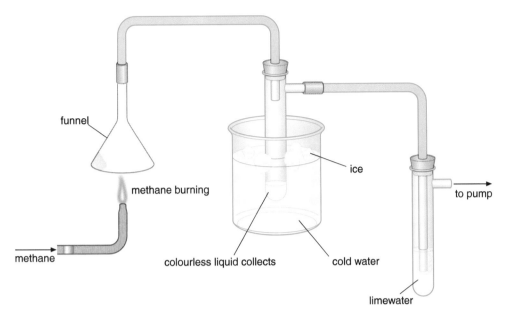

funnel

methane burning

methane

colourless liquid collects

ice

cold water

to pump

limewater

◀ *Burning methane in oxygen*

The equation for the complete combustion of ethane is shown below:

$$2C_2H_6 + 7O_2 \rightarrow 4CO_2 + 6H_2O$$

e How many molecules of oxygen are needed to burn 1 molecule of ethane completely?

Not enough oxygen

If a fuel is burned without enough oxygen then **incomplete combustion** takes place and **carbon monoxide** is produced. This is a poisonous gas. Carbon (soot) is also produced. The incomplete combustion of methane can be shown by the following word equation:

methane + oxygen → carbon monoxide + carbon + water.

f Write the balanced symbol equation for the incomplete combustion of ethane, C_2H_6.

It is better for hydrocarbon fuels to burn with complete combustion than incomplete combustion because:
• less soot is produced
• more heat is produced
• carbon monoxide is not produced.

Bunsen flames

The photographs on the right show two Bunsen burner flames. The top is the blue flame and the bottom one is the yellow flame.

When we want to heat something to a high temperature we use the blue flame. The blue flame gives out more energy than the yellow flame. This is because it involves complete combustion.

The blue flame is also a cleaner flame because it produces less soot than the yellow flame which produces lots of soot because it involves incomplete combustion.

Death by carbon monoxide

Girl Dies of Carbon Monoxide Poisoning

A girl, aged nineteen, was found dead yesterday in her flat. It is believed she died of carbon monoxide poisoning. Scientists have taken away a gas fire for examination.

Every year in the UK, 50 people die of carbon monoxide poisoning. Carbon monoxide is a poisonous gas. It forms when incomplete combustion of a fuel takes place. Carbon monoxide is colourless and has no smell or taste. People die without realising they are breathing in carbon monoxide. You can now buy carbon monoxide detectors in DIY stores, but few people have them.

Carbon monoxide forms a compound called carboxyhaemoglobin with the red cells in the blood. This prevents oxygen circulating the body in the blood.

▲ *A car having its MOT test*

The equation shows the incomplete combustion of methane.

$$4CH_4 + 5O_2 \rightarrow 2CO + 2C + 8H_2O$$

methane + oxygen → carbon monoxide + water

Carbon monoxide usually forms when there is not enough ventilation for the fuel to burn completely. Servicing gas boilers and fires each year removes soot and ensures enough air can enter. This should prevent carbon monoxide from forming.

Car engines produce carbon monoxide. There are limits to how much carbon monoxide a car can give out. This is measured as part of the MOT test. Cars exceeding the limit fail the test.

Questions

1 Why is carbon monoxide such a dangerous gas?

2 Suggest ways in which the chances of carbon monoxide poisoning happening in a house can be minimised.

3 How many molecules of oxygen are needed for the incomplete combustion of one molecule of methane?

4 In long road tunnels there is sometimes a sign warning drivers to switch off their engines if they are stationary. Why?

5 Suggest one other reason why regular servicing of gas boilers should be done, apart from avoiding poisoning.

Feeling energetic

In this item you will find out

- about exothermic and endothermic reactions

- how reactions can be understood in terms of bond making and bond breaking

- how to calculate the energy output of a fuel

Fuels are materials that can release energy. Usually this is done by burning the fuel in air or oxygen.

In this forest fire, the trees are the fuel. This fire produces a huge amount of energy.

Different fuels produce different amounts of energy.

The table gives the energy released when equivalent quantities of different fuels are burned.

Food additive	Hexane	Methane	Methanol	Carbon	Hydrogen
State of fuel at room temperature	liquid	gas	liquid	solid	gas
Energy released (kJ)	4163	890	726	393	286

a Which liquid fuel in the table releases most energy?

b Suggest one advantage of burning methane to produce electricity rather than coal (carbon).

An **exothermic** reaction is a chemical reaction giving out energy. All of the materials in the table undergo exothermic reactions when they burn. Energy is transferred to the surroundings.

c Can you think of other exothermic reactions?

An **endothermic** reaction is a chemical reaction taking in energy from the surroundings.

Amazing fact

There are very few endothermic reactions and very many exothermic ones. At one time, people thought that endothermic reactions were impossible.

Recognising reactions

Sam carries out an experiment. He reacts water with solid calcium oxide to produce calcium hydroxide. He puts a thermometer into some water in a test tube. Then he adds a few small pieces of calcium oxide to the water. When he shakes the test tube, the temperature rises. This is an exothermic reaction because energy is given out to heat up the water. An endothermic reaction would produce a decrease in temperature.

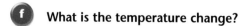

cover

ethanol

spirit lamp

top pan balance

Comparing energy changes in combustion reactions

Combustion reactions are reactions where a fuel burns in oxygen to release energy.

Alcohols, such as methanol and ethanol, burn easily in small spirit burners.

The equation for the complete combustion of ethanol is:

$$C_2H_5OH + 3O_2 \rightarrow 2CO_2 + 3H_2O$$

The diagram on the left shows a small spirit lamp containing ethanol being weighed before the experiment.

d Why is the spirit lamp covered?

The lit spirit lamp is placed under a metal can called a **calorimeter** filled with 100 g of water. The temperature of the water is measured.

Look at the apparatus in the diagram below.

As the ethanol burns, heat energy is transferred to the water.

e What happens to the temperature of the water as the ethanol burns?

At the end of the experiment, the flame is extinguished and the spirit lamp is reweighed.

The temperature of the water at the end of the experiment is recorded.

Sample results:
Temperature of the water
at the start = 20 °C
Temperature of the water
at the end = 32 °C
Mass of spirit lamp
at the start = 105.57 g
Mass of spirit lamp
at the end = 103.47 g

f What is the temperature change?

g What mass of ethanol has burned?

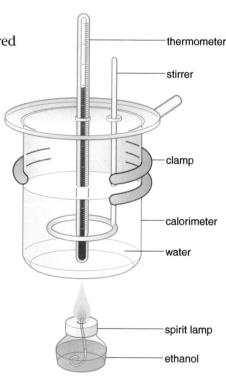

thermometer

stirrer

clamp

calorimeter

water

spirit lamp

ethanol

▲ Set up for calorimetry experiment

The amount of heat energy taken in by the water in the can is calculated by the equation:

Energy = mass (in g) × specific heat capacity of water × temperature rise (in °C)

The specific heat capacity of water is 4.2J/g/°C.

The answer is given in joules (J). This can be converted into kilojoules (kJ) by dividing by 1000.

 Show that the energy taken in by the water in the calorimeter is 50.4kJ.

Energy transfer

We assume that all the energy produced when the ethanol burns goes to heat up the water in the can.

A useful comparison of the energy value of fuels, is the energy given out per gram. The equation is:

$$\text{energy per gram (in J/g)} = \frac{\text{energy supplied (in J)}}{\text{mass of fuel burned (in g)}}$$

i **Calculate the energy given out by burning 1g of ethanol.**

The value from this experiment is compared with the value in a data book. The value in the data book is much higher.

j **Suggest reasons why the values are different.**

By convention, the energy change in an exothermic reaction is given a negative sign.

The energy per gram value in the data book is −29.7kJ/g or −29700J/g.

Bond breaking and bond making.

The complete combustion of ethanol can be shown using displayed formulas:

$$H-\underset{\underset{H}{|}}{\overset{\overset{H}{|}}{C}}-\underset{\underset{H}{|}}{\overset{\overset{H}{|}}{C}}-O-H + 3\,O{=}O \longrightarrow 2\,O{=}C{=}O + 3\,\overset{H}{\underset{H}{O}}$$

Each of the bonds is a covalent bond. During the reaction, some bonds have to be broken and some new bonds have to be formed. Breaking a bond requires energy: the process is endothermic. Forming new bonds gives out energy: the process is exothermic.

In this reaction:

5 C—H bonds, 1 C—C bond, 1 C—O bonds, 1 O—H bond and 3 O=O bonds are broken.

4 C=O bonds and 6 O—H bonds are formed.

More energy is given out when the new bonds are formed, than when the old bonds are broken. Therefore, the reaction is exothermic.

Examiner's tip

Units are important. A missing or wrong unit may cost you a mark.

keywords

calorimeter • endothermic • exothermic

Coffee on the run

Steve and Leah enjoy hill walking. Usually when they get thirsty, they stop for a cup of coffee which they bring with them in a flask. The flask is quite heavy so it is a nuisance to carry around sometimes.

Steve is reading a walking magazine one day when he notices an advert for a new coffee product which has just been introduced. The next time he meets Leah he tells her about it.

'This new product is ideal for people who go out for the day into the countryside but like a cup of hot coffee.

All you have to do to get hot coffee is to press the plastic button at the bottom of the can. This starts a chemical reaction inside the can. The reaction is exothermic and the energy from this reaction heats the coffee in the can.

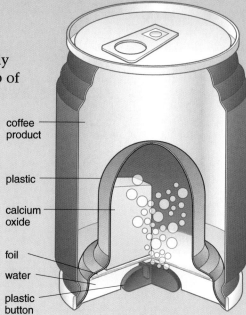

coffee product

plastic

calcium oxide

foil

water

plastic button

▲ *Inside the self-heating can*

The chemical reaction that heats the coffee takes place between calcium oxide and water. They are stored separately at the bottom of the can. When the plastic button is pressed, water escapes and comes into contact with the calcium oxide.

Look, here is a diagram of it. The energy released by this reaction heats up the coffee.'

Questions

1 Suggest another way of getting a hot cup of coffee if this kind of product was not available.

2 Why is it important that the calcium oxide and water are stored separately?

3 Why is it important that the coffee does not come into contact with the chemicals?

4 The equation for the reaction is: $CaO + H_2O \rightarrow Ca(OH)_2$
What is the name of the product of the reaction?

5 How many atoms are there in the formula of the product?

6 The same product is formed when calcium reacts with water. Hydrogen, H_2, is also produced. Write a balanced symbol equation for the reaction of calcium and water.

7 The reaction of calcium and water is also exothermic. Suggest why this reaction would be unsuitable for heating the coffee can.

8 In trials of the self-heating can, the coffee either did not get hot enough or got too hot. What could be changed to improve this?

C1a

1 Describe the chemical test for carbon dioxide? [2]

2 **a** Finish the word equation for the decomposition of sodium hydrogencarbonate.

sodium hydrogencarbonate → ____(1) + ____(2) + ____(3) [3]

b Sodium hydrogencarbonate is heated in test tube. Does its mass increase, decrease or stay the same? Explain why. [2]

3 The formula of sodium hydrogencarbonate is $NaHCO_3$. There are four elements combined. How many atoms of each element are in the formula? [2]

4 Finish the symbol equation for the decomposition of sodium hydrogencarbonate.

$NaHCO_3$ → [4]

5 Ammonium hydrogencarbonate, NH_4HCO_3, is used as a raising agent in biscuits. On heating it decomposes to give ammonia gas (NH_3), water and carbon dioxide.

a Write a symbol equation for the decomposition of ammonium hydrogencarbonate. [4]

b Suggest a benefit of using ammonium hydrogencarbonate rather than sodium hydrogencarbonate. [2]

6 Describe and explain the changes that take place when a potato is boiled. [2]

C1b

1 What are the special features of an emulsifier molecule? [2]

2 Foods contain antioxidants.

a What does an antioxidant do in a food? [1]

b Give two foods that contain antioxidants. [2]

3 Explain why removal of water from food slows down the spoiling of food. [2]

4 Describe how an emulsifier helps to prevent oil and water separating in an emulsion. [3]

C1c

1 Ethanoic acid, CH_3CO_2H, and propanol, $CH_3CH_2CH_2OH$, react together to form an ester.

a What conditions are needed to form the ester? [2]

b What is the name of the ester? [1]

c Draw the displayed formula of this ester. [2]

2 Explain why water is not a solvent for nail varnish. Use ideas about particles in your answer. [3]

C1d

1 Arrange the four fractions in order of increasing boiling point.

fuel oil   heating oil   paraffin   petrol [3]

2 The table contains the boiling temperature range for four fractions from the fractional distillation column.

Fraction	Boiling temperature range in °C
A	70–120
B	120–170
C	170–220
D	220–270

a Which fraction contains the largest molecules? [1]

b Which fraction comes out at the highest point in the column? [1]

3 A hydrocarbon has a formula $C_{12}H_{26}$. Cracking breaks this hydrocarbon into small saturated and unsaturated products.

a Write a symbol equation for the two products each containing two carbon atoms. [3]

b Write down the names of these products. [2]

C1e

1 The linking of small ethene molecules together to form long chains is called ____. [1]

2 The displayed formula for propene is shown below.

Copy the diagram and draw a ring around the feature of the molecule that is typical of all alkenes. [1]

3 The equation for the reaction of propene and hydrogen is:

$$C_3H_6 + H_2 \rightarrow C_3H_8$$

  a Draw the displayed formula and name the product. [2]

  b What type of reaction is this? [1]

4 The diagram shows part of a polymer chain.

  a Why is this polymer an addition polymer? [1]

  b Draw the displayed formula of monomer. [1]

5 The displayed formula of tetrafluoroethene is:

Draw the displayed formula of the polymer poly(tetrafluoroethene). [2]

6 Describe how you could show that an alkene is unsaturated. [2]

C1f

1 Most polymers are not biodegradable. Suggest two disadvantages of dumping polymer waste in landfill sites. [2]

2 The table gives information about three materials.

Material	Absorbs water	Sweat absorbed	Breathable
cotton	yes	absorbed	no
nylon	no	not absorbed	no
Gore-Tex®	no	escapes through material	yes

Which material is best for making an anorak for a mountain climber? Explain your choice. [2]

3 Explain why Gore-Tex® lets perspiration out but does not let water in. [2]

4 The diagrams show parts of two polymer chains.

        A   B

  a The force holding two M units together in A is a ____ bond [1]

  b The force between two chains in A is an ____ force. [1]

  c The polymer B has ____ between chains. [1]

  d Why does polymer A have a low melting point and can it be easily be stretched? [1]

  e What are the characteristic properties of polymer B? [2]

C1g

1 Landlords renting flats to students must have a certificate each year to show that all gas appliances have been serviced. Why is this? [2]

2 A power station in East Anglia is using straw as a fuel to produce electricity. The straw is left over when wheat is grown. Suggest an advantage and a disadvantage of using straw as a fuel. [2]

3 Write word equations for complete and incomplete combustion of propane. [4]

4 Explain why every year the consumption of fossil fuels in the world rises. [2]

5 Write balanced symbol equations for:

  a the complete combustion of propane, C_3H_8 [2]

  b the incomplete combustion of propane, C_3H_8 [2]

C1h

1 In an experiment to burn a small sample of fuel weighing 0.1 g, the energy given out is 20 000 J.

How much energy would be given out if 1 g of same fuel is burned?

A 2000 kJ

B 20 000 J

C 200 000 J

D 2 000 000 J [1]

2 *The word equation for the process of photosynthesis is shown below.*

carbon dioxide + water + energy → glucose + oxygen

  a *Is this reaction endothermic or exothermic? Explain your answer.* [2]
  b *Finish the word equation for respiration producing carbon dioxide and water. Include '+ energy' in your equation.* [2]

3 *The graph shows the energy produced when 1 formula mass of different alcohols is burned. These values are taken from a textbook.*

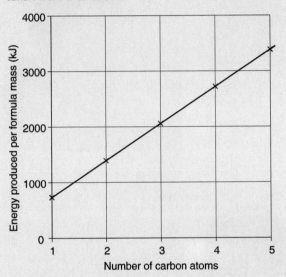

  a *What pattern is there in the results?* [1]
  b *Copy the graph and sketch the line you would expect from experimental results.* [1]
  c *The diagram shows apparatus that gives better results.*

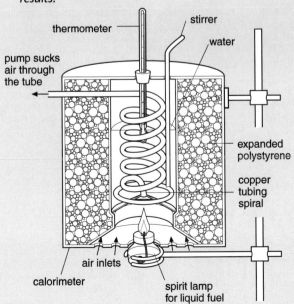

Explain why the results using this apparatus are better. [2]

4 *The table gives theoretical values for the energy produced when 1 formula mass of different alcohols is burned.*

Alcohol	Formula	Formula mass	Energy produced (kJ)
methanol	CH_4O	32	–726
ethanol	C_2H_6O	46	–1367
propanol	C_3H_8O	60	–2021
butanol	$C_4H_{10}O$		–2675
pentanol		88	–3329

  a *What are the two items missing from the table?* [2]
  b *What is the trend in the amount of energy produced **per gram** when these alcohols burn?* [2]

5 *The diagram summarises the bonding changes that take place when hydrogen and chlorine react to form hydrogen chloride.*

$$H—H \quad Cl—Cl \quad → \quad H—Cl \quad H—Cl$$

Explain why the reaction is exothermic. Use ideas of bond breaking and bond forming in your answer. [3]

6 *In an experiment to find the energy given out when 0.25 g of X burns, the following results are obtained:*

Mass of water = 100 g
Specific heat capacity of water = 4.2 J/g/°C
Temperature changes from 22 to 28°C.

  a *Calculate the heat energy received by the water.* [3]
  b *Calculate the energy given out per gram.* [1]

7 *Complete combustion of 32 g of methanol produces 726 kJ of energy.*

  a *Calculate the energy produced per gram when methanol burns completely. (Look back to question 4.)* [2]
  b *Why is it better to compare the energy released when one formula mass of each alcohol is burned, rather than the energy released when 1 g of each alcohol is burned?* [1]
  c *Write a balanced symbol equation for the complete combustion of methanol.* [3]
  d *Describe which bonds, and how many, are broken. Which bonds are formed when methanol burns completely?* [1]

C2 Rocks and metals

Drivers now have to pay £8 to drive their cars into London each day. Is this to reduce atmospheric pollution in London?

A steel bridge has legs dipping into the sea. How do they stop them rusting? They can't paint them.

Are natural disasters, such as earthquakes, occurring more often or do we just hear more in the media?

- All too regularly we hear of natural disasters such as earthquakes, tsunamis and hurricanes.

- Can we accurately predict where the next disaster will happen so we can take some precautions? The answer is probably no, because we do not really know enough about the structure of the Earth.

- Japan suffers from earthquakes as it is at the junction of three of the large plates that make up the outer crust of the Earth. The Japanese are drilling deep into the Earth – three times deeper than anyone before, in order to find out more.

- This research, and other research taking place around the world, may help us to predict natural disasters in the future.

What you need to know

- The possible effects of burning fossil fuels on the environment.

- The way rocks are formed and destroyed.

- Air is a mixture of gases.

Colour the world

In this item you will find out

- about different types of pigments and paint

- about dyes for fabrics

Do you like to wear colourful clothes? People wear much more colourful and brighter clothes today than they did two hundred years ago.

The dyes available to colour fabrics a couple of hundred years ago were very limited. The only dyes were natural dyes from plants and animal materials. For example, fabric was dyed yellow by boiling with water containing onion skins. These natural dyes were dull compared to modern synthetic dyes and tended to wash out of clothes.

Synthetic dyes were discovered about 130 years ago. This discovery revolutionised the dyeing industry. Clothes could now be produced which were colourfast and had a wide variety of bright colours.

The first synthetic dye was made by William Perkin. It is called mauvine and is a rich purple colour. It was made from coal tar.

Today almost all dyes used in industry are synthetic dyes.

 a Apart from clothes, what are dyes used for today?

Amazing fact

It is hard to believe that the source of many brightly coloured dyes is black coal tar.

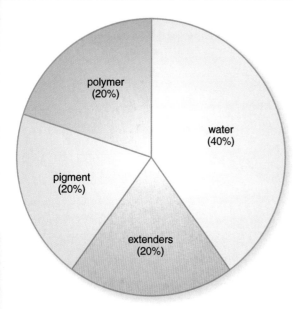

▲ *Composition of water-based paint*

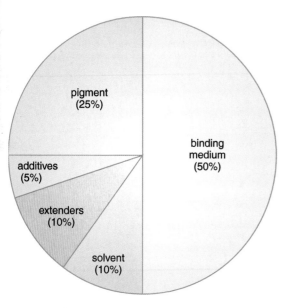

▲ *Composition of oil-based paint*

Paint particles

All paints contain particles of coloured pigments. They also contain a solvent and a binding medium. The solvent thins the paint while the binding medium hardens as it reacts with oxygen to form a layer of paint.

Paint is a mixture called a **colloid**. This is where solid particles are mixed with liquid particles but are not dissolved. The pigment particles and binding medium particles are dispersed throughout the liquid solvent particles. The different component particles of the paint do not separate out because they are scattered throughout the mixture. The particles are also small enough so that they do not settle at the bottom.

Water or oil?

Most paints are applied as a thin surface which dries when the solvent evaporates. Different paints have different solvents.

Emulsion paints are water-based paints. They consist of tiny drops of a liquid polymer (the binding medium) spread out in water (the solvent). After the paint is applied, the water in the paint evaporates and the polymer particles fuse together to form a continuous film.

b How would you clean paint brushes used for emulsion paint?

Oil paints use hydrocarbon oil as the solvent and consist of pigments spread throughout the oil. They often contain an extra solvent that dissolves the oil to form a solution. After the paint is applied, the solvent evaporates and the binding medium reacts with oxygen from the air (oxidises) to form a paint film.

Gloss paints used to be made from a hardening oil such as linseed oil. These harden to form a thin layer by reaction with oxygen. Linseed oil is still used to treat cricket bats. This hardening, however, is very slow taking several days. We like paints to dry to a hard film quickly. A drying time of a couple of hours is preferable.

Brushes used with linseed oil paints are more difficult to clean.

Modern gloss paints used an alkyd resin binding medium. The alkyd resin is a polymer formed by reacting a vegetable oil, such as linseed oil, with an alcohol and an organic acid. Brushes used for modern oil paints are easier to clean providing the paint is not allowed to dry.

c Suggest why gloss paint should be used in a well-ventilated room?

d Suggest why is it better to have several thin coats of paint rather than one thick coat.

Special paints

There are now special paints available. Some paints contain special pigments whose colour changes as the temperature changes. These pigments are called **thermochromic pigments** and the paints are called thermochromic paints. The paints could be used to coat a cup, for example. The colour of the cup will change when hot liquid is put into it.

e   **Why would a cup that changes colour be an advantage?**

Thermochromic paints could also be used on the outside of a kettle. The paint would change colour when the water inside the kettle was hot.

f   **Why would this be a safety feature of the kettle?**

Thermochromic paints can also be added to acrylic paints to give even more colour changes.

There are other paints that glow in the dark. They contain **phosphorescent pigments**. These pigments absorb and store light energy and then release it as light over several hours. This means you can see the pigments in the dark.

Beer is best drunk cold. A new brand of beer is sold in special cans. The cans are labelled by printing using an ink containing a thermochromic pigment. At ordinary temperatures the labelling is red, but if the can is cooled to the correct drinking temperature the printing goes blue. A blue can is ready for drinking.

Before phosphorescent pigments were used for such purposes, radioactive materials used to be used on watch faces. If a Geiger counter was passed over the face of one of these watches there would be a rapid clicking sound showing radiation was being detected. The amounts of radiation were small but significant.

Modern paint technologists are striving to produce new paints for specific purposes.

You can buy paints with special finishes e.g. a paint with a suede finish or paints for special surfaces e.g. for bathroom tiles or for hot surfaces. Paint technologists have to modify the composition of paints and test them thoroughly before they can be sold.

g   **Why would these paints be useful for seeing road signs at night on country roads?**

h   **Suggest why phosphorescent paints are safer than paints containing radioactive substances which also glow in the dark.**

i   **Why would phosphorescent paint be unsuitable for a sign warning people inside a dark underground cavern about a possible danger?**

keywords

colloid • phosphorescent pigment • thermochromic pigment

Eco-friendly paint

Shari and Ahmed are decorating a nursery for their baby which is due in March. Shari wants to paint the room in bright colours. They pay a visit to a local DIY store to look at paint. They pick up some tins of gloss paint from a well-known brand.

'I'm worried about what's in these paints,' says Shari, 'Babies are more likely to be poisoned by the chemicals in paint than adults.'

They pick up a leaflet which explains what is in the gloss paint.

'I think this contains too many chemicals,' says Shari.

She asks for help in the store and the assistant explains that when the paints have dried the resulting paint film is tough and hard.

A couple of days later, Shari sees a website on the Internet which is advertising eco-friendly gloss paints. They contain natural ingredients such as plant oil, earth pigments and wood resin. Shari notices that the range of colours is smaller than for normal gloss paints and they are less bright. The eco-friendly paints are biodegradable. This means they will rot down when they are disposed of.

GLOSS PAINT

Hydrocarbon solvent
Acrylic pigment
Alkyd resin binder
Extenders
Additives to speed up drying

▲ *Normal gloss paint*

GLOSS PAINT

Plant oil
Earth pigments
Wood resin
Extenders

▲ *Eco-friendly gloss paint*

Questions

1 Why is Shari concerned about the ingredients in normal gloss paint?

2 Which paint would dry faster? Explain your answer.

3 Suggest why the eco-friendly paint comes in fewer colours than normal gloss paints?

Building basics

In this item you will find out

- about materials used in construction

- about sedimentary, metamorphic and igneous rocks

- how limestone can be used to make cement and concrete

The photograph on the right shows the tallest skyscraper in the world. It is the Taipei 101 tower in Taiwan. It is over 508 m high. This is more than ten times the height of Nelson's column in London.

This skyscraper is made mainly from steel-reinforced concrete and glass. These **construction materials** are made from other materials taken from the earth.

Steel is made from iron which is extracted from **ores**. Aluminium, which is another popular building material, is also extracted from ores. Glass is made by heating sand to very high temperatures.

The skyscraper also contains bricks. They are made from baked clay.

Concrete is made from **limestone**. Limestone used to be an important building material.

▲ *The Taipei 101 tower in Taiwan*

The wall in the photograph is in the Peak District in Derbyshire. It was made from pieces of limestone rock from a nearby quarry. Today, most limestone is used to make other construction materials.

a Suggest one advantage of using rocks close to where they are quarried.

Amazing fact

The Taipei 101 tower is built to resist an earthquake of 7 on the Richter scale. During construction it withstood an earthquake of 6.8 successfully.

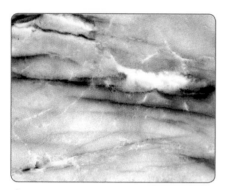
▲ Limestone

▲ Marble

▲ Granite

Granite, limestone and marble

The photographs show samples of **granite**, limestone and **marble**. They are three rocks used as construction materials.

Limestone is a sedimentary rock. It is formed when small sea creatures die, sink to the bottom of the ocean and are buried. Over a long period of time the rock is formed. The remains of the sea creatures from millions of years ago may be seen as fossils in limestone today.

Marble is a metamorphic rock. It is made by the action of high pressures and high temperatures on limestone. During the process, limestone melts and crystallises.

Granite is an **igneous** rock. It is a crystalline rock formed when molten rock from inside the Earth cools and crystallises within the Earth.

Granite is much harder than marble. Marble is much harder than limestone. In limestone the tiny grains are cemented together but can be separated quite easily. In marble there is some fusion of the grains together under the effects of high pressures and temperatures, making it harder. Granite is hardest because it is crystalline.

Decomposition of limestone

Limestone and marble are both forms of calcium carbonate, $CaCO_3$.

Thermal decomposition is a reaction where one substance is chemically changed, by heating, into at least two new substances. Calcium carbonate splits up (thermally decomposes) to form calcium oxide and carbon dioxide when it is heated.

calcium carbonate → calcium oxide + carbon dioxide

$$CaCO_3 \rightarrow CaO + CO_2$$

The process of heating limestone to produce calcium oxide (sometimes called quicklime) and carbon dioxide goes back many centuries. Originally it was done in large kilns heated by burning wood. These kilns were usually covered with earth. They produced calcium oxide in batches. The kiln had to be emptied and refilled after each batch.

Later, stone kilns using coal as the fuel were developed which enabled calcium oxide to be produced continuously. Today, modern gas-fired kilns are used.

b The kilns need to get temperatures above 900 °C. Suggest a reason for covering the kiln with earth.

c What is the advantage of a continuous process over a batch process?

Lime is still important for manufacturing sodium carbonate, steel manufacture, water purification and treatment. Over half of the output in the UK comes from the area around Buxton in Derbyshire.

Cement and concrete

A lot of limestone is made into **cement**. To do this limestone and clay are heated together. The product is then crushed. When cement is mixed with sand and water, it sets hard. It can be used to stick bricks together when building.

Cement can be used to make concrete. Cement, sand, gravel (small stones) and water are mixed together to make concrete.

Concrete is a very useful construction material for making railway sleepers or lamp posts. It is not very strong, however, unless it is reinforced.

Reinforcing concrete makes a **composite** material. Reinforcing the concrete with steel rods or a steel framework makes concrete a better construction material. It combines the hardness of concrete with the flexibility of steel. The forces are transferred sideways which makes the beam stronger. In the diagram below the vertical force is turned into a horizontal one.

The photograph above shows concrete bridges where steel is used to reinforce the concrete.

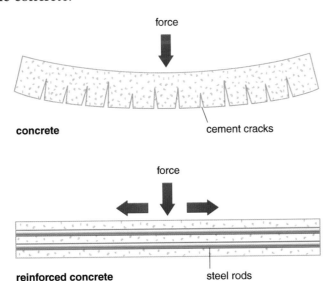

force

concrete   cement cracks

force

reinforced concrete   steel rods

▲ Steel rods make concrete stronger

keywords

cement • composite • concrete • construction materials • granite • igneous • limestone • marble • ores • thermal decomposition

Uses of limestone

Every year 100 millions tonnes of limestone are quarried in the UK.

The pie diagram below shows the main uses of limestone.

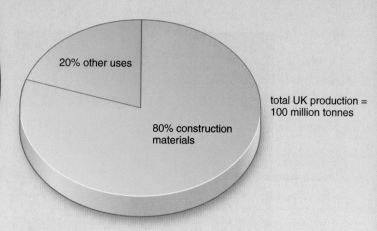

20% other uses

80% construction materials

total UK production = 100 million tonnes

Much of the limestone used as a construction material is known as aggregate. It is limestone that is crushed into small pieces. This is used where solid foundations are needed, for example for concrete floors, roads or railway tracks.

The second pie diagram shows how the limestone not used as a construction material is used.

A lot of the limestone is converted to calcium oxide and carbon dioxide by thermal decomposition. Calcium oxide is sometimes called quicklime. If calcium oxide is reacted with water, calcium hydroxide, $Ca(OH)_2$, is formed. This is sometimes called slaked lime. Farmers use calcium hydroxide to neutralise the soil.

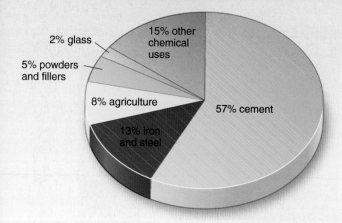

2% glass

5% powders and fillers

8% agriculture

13% iron and steel

15% other chemical uses

57% cement

Questions

1 How many atoms are there of each element in the formula of calcium hydroxide.

2 Suggest why is it easier to transport the aggregate than large lumps of limestone?

3 What mass of limestone is turned into cement each year in the UK? Use the two pie charts to help you.

4 Suggest why the mass of calcium oxide produced by heating calcium carbonate is always less than the mass of calcium carbonate used.

5 Write a balanced equation for the reaction of calcium oxide and water.

6 Calcium hydroxide can be used to make lime mortar. When this sets it reacts with carbon dioxide from the air. The product is calcium carbonate. Write a symbol equation for this reaction.

Restless Earth

In this item you will find out

- about the structure of the Earth

- what causes earthquakes and volcanoes

- about volcanic rocks

A volcanic eruption can cause widespread damage. Geologists study **volcanoes** for several reasons.

They carry out experiments to find out more about what is happening inside a volcano. The results of their experiments may reveal information about what is inside the Earth and also enable them to make predictions about when an eruption of a volcano may take place.

It is also possible to monitor, with sensitive equipment, the areas where earthquakes are likely to happen. Slight changes within the Earth can be detected.

This kind of monitoring of possible earthquake and volcano sites is very expensive and needs to be carried out over long periods of time if useful predictions are to be made.

a Suggest why there is much more monitoring in the US than in Indonesia.

To understand these natural phenomena better we need to know more about the Earth and its structure. But there are problems with studying the structure of the Earth. We cannot just slice it open and look inside. The Earth has a diameter of about 12,700 km. The deepest that we have been able to drill into the Earth is 15 km.

▶ Earthquakes can cause devastation and loss of life

Amazing fact

On the small island of Bali in Indonesia there are 149 active volcanoes.

Structure of the Earth

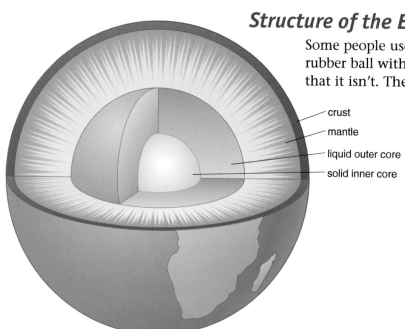

crust
mantle
liquid outer core
solid inner core

Some people used to think that the Earth was like a solid rubber ball with the same material throughout. We now know that it isn't. The diagram shows the structure of the Earth.

The Earth consists of a **core**, **mantle** and rocky **crust**. The crust is between 10 km thick under the oceans and 65 km under the land. The mantle is the zone between the core and the crust. The mantle is relatively cold and rigid just below the crust. At greater depths it is hot and non-rigid. At these levels rocks can move and flow and **convection currents** can be set up.

The outer 100 km of the Earth, consisting of the crust and the upper part of the mantle, is called the **lithosphere**. It is relatively cold and rigid.

 b   Where would be the best place to drill down into the crust to reach the mantle?

Tectonic plates and earthquakes

The lithosphere is made up of a number of large sections called **tectonic plates**. There are two types of tectonic plates: oceanic and continental. The oceanic plates lie under the oceans and the continental plates form the continents. These plates float on the mantle because they are less dense than it. They move a few centimetres each year.

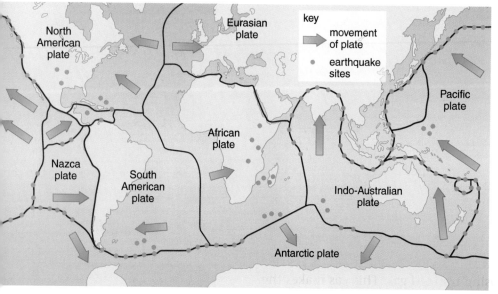

North American plate

Eurasian plate

key

→ movement of plate

• earthquake sites

Nazca plate

South American plate

African plate

Pacific plate

Indo-Australian plate

Antarctic plate

The diagram shows the plates that cover the Earth.

You will notice the orange dots on the map. These represent places where earthquakes occur.

c   Can you suggest a theory about where earthquakes occur?

d   Does this theory fit all cases?

Earthquakes occur where two plates join. The sliding of one plate against the other builds up stresses and strains. When these become too much the result is an earthquake.

Theory of plate tectonics

When two plates collide, the rocks are squeezed together. This is shown in the diagram. As the two plates collide the oceanic plate is forced under the continental plate. The rocks in the oceanic plate partly melt and return to the **magma** (molten rock) in the mantle. This is called **subduction**. The result is an oceanic trench, where the oceanic plate dips, and a mountain chain caused by folding of rocks on the continental plate.

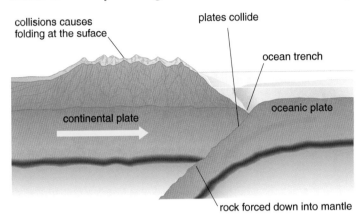

The continental plate has an average density of 2.7 g/cm^3. The oceanic plate is more dense and has a density of 3.3 g/cm^3. The convection currents caused by energy transfer in the mantle cause the plates to move slowly.

▲ *Rhyolite*

Volcanic rock

Where there are weaknesses in the Earth's crust such as cracks or volcanoes, molten rock can make its way through to the surface. The magma rises to the surface, usually because it is less dense than the mantle. But if there is enough pressure the magma can rise even if it is denser than the mantle.

Magma cools and crystallises on the surface to form igneous rock. The magma can have different compositions and this affects what types of rock are formed.

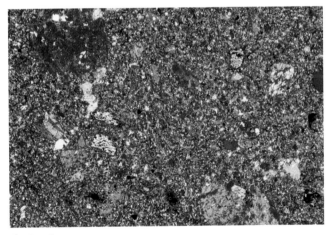

▲ *Basalt*

Lava with a low silica content is runny. It flows steadily so its movement is predictable and fairly 'safe'. It quickly cools forming iron-rich rocks like **basalt**, which have only very tiny crystals.

A high silica content prevents lava flowing easily so it piles up forming rocks like **rhyolite** (pumice stone), which is used as an abrasive. High silica content also stops lava from losing trapped gas. This gas makes the lava explode out of volcanoes as dangerous and unpredictable lava bombs.

Igneous rocks can also cool slowly inside the Earth. Here, very slow cooling allows large interlocking crystals to form in rocks such as granite. Slightly faster cooling gives smaller interlocking crystals in rocks such as gabbro.

keywords

basalt • convection currents • core • crust • lithosphere • magma • mantle • rhyolite • subduction • tectonic plates • volcano

Developing the theory of plate tectonics

The theory of plate tectonics on page 121 has been developed fairly recently. In 1915, a German geophysicist, Alfred Wegener, published a book. He was one of the first people to suggest that continents moved. He suggested that a supercontinent, which he called *Pangaea*, had existed. He thought that it had broken up, starting 200 million years ago, and that the pieces had 'drifted' to their present positions.

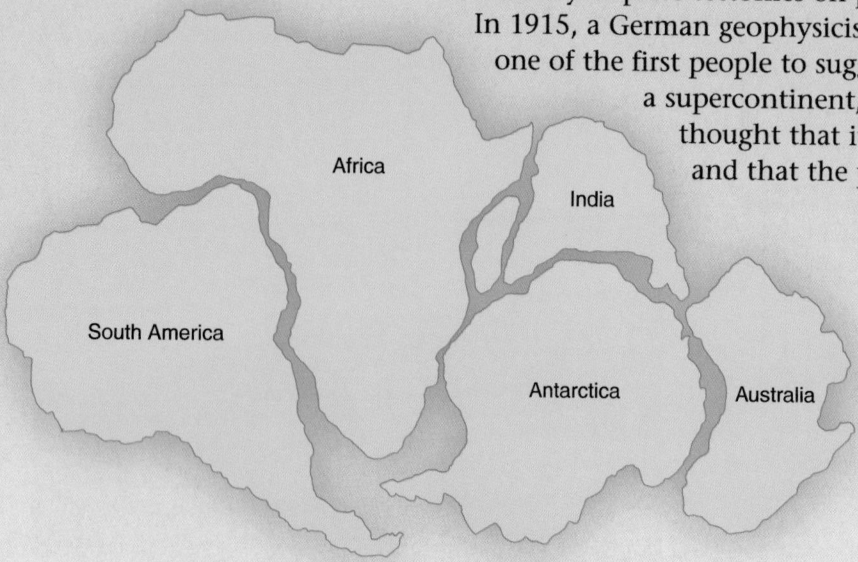

The diagram on the left shows *Pangea*.

The book was not translated into English until 1924 when it was criticised by scientists. The president of the prestigious American Philosophical Society said of Wegener's ideas. 'Utter, damned rot!' A leading British geologist said, 'Anyone who valued his reputation for scientific sanity would never dare support such a theory.'

If Wegener's ideas were true, South America was once joined to Africa. He also noted that when you fit Africa and South America together, mountain ranges (and coal deposits) run uninterrupted across both continents.

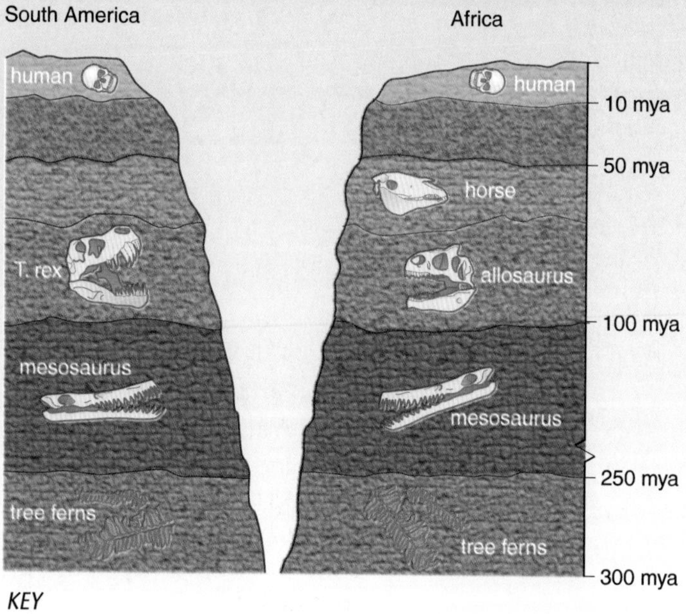

Wegener compared the fossils from Africa and South America. These are shown in the diagram.

The problem with Wegener's hypothesis was that he could not provide evidence to support his theories or explain what caused the movement of the plates. The rotation of the Earth and tidal-type waves were two ideas that did not seem credible.

By 1954 scientists realised that there were convection currents in the mantle inside the Earth. It was possible to make much more accurate measurements, including measurements of magnetic fields. By about 1960, scientists generally agreed with Wegener's theory. They realised that Alfred Wegener was right in most of his major ideas.

KEY
mya = million years ago

Questions

1 Alfred Wegener was a geophysicist. Which sciences does a geophysicist specialise in?

2 What does the fossil evidence in the diagram above suggest?

3 Suggest why there was opposition to his ideas in 1924?

4 Why was there little opposition by 1960?

Make mine metals

In this item you will find out

- how copper is purified
- about alloys
- the use of smart alloys

The photograph shows some objects that are made of the metal copper. At the front of the picture on the left is malachite. This is an ore of copper.

Copper is an expensive metal although it is relatively easy to extract from its ore. It is expensive for the following reasons:

- There is a huge demand for copper for electrical wires and water pipes.
- There are very limited stocks of ores, such as malachite, in the Earth.
- Most of these ores contain only a small percentage of copper.

a Explain why it is worth using ores containing a very low percentage of copper.

Recycling copper from electrical wires, water pipes and **alloys**, for example, is cheaper than extracting copper from copper ore. Recycling also makes our scarce resources last longer. But local councils arrange for the collection of a range of materials, such as glass and paper, for recycling. They do not collect copper even though waste copper is much more valuable than waste glass or paper.

b Suggest reasons why they do not collect waste copper.

Amazing fact

'Silver' coins such as 50p pieces contain a lot of copper but no silver. The coins are made from an alloy of copper and nickel.

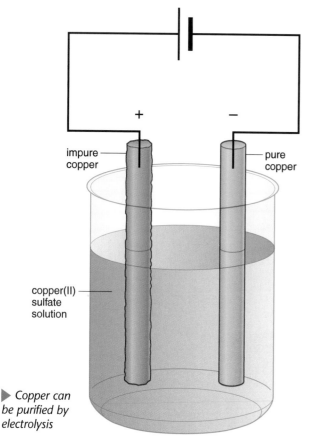

▶ Copper can be purified by electrolysis

impure copper

pure copper

copper(II) sulfate solution

◀ Purification of copper on a large scale

Purification of copper

Copper is easily extracted from its ore by heating the ore with carbon. This is because it is low in the reactivity series (page 130). Its ores are unstable and split up easily by reduction.

Copper must be very pure if it is to be used to make electricity wires – at least 99.95% pure. Any impurities will increase the resistance of the wire.

Copper can be purified by **electrolysis**. The apparatus is shown in the diagram on the left.

The copper(II) sulfate solution is the **electrolyte** and contains:

copper(II) ions   sulfate ions   hydrogen ions   hydroxide ions

Cu^{2+} SO_4^{2-} H^+ OH^-

c The copper(II) ions and the sulfate ions come from the copper(II) sulfate. Where do the hydrogen and hydroxide ions come from?

During the electrolysis, copper atoms from the positive electrode (**anode**) lose electrons and form copper(II) ions in solution.

copper atom → copper(II) ion + 2 electrons

$$Cu \quad \rightarrow \quad Cu^{2+} \quad + \quad 2e^-$$

At the negative electrode (**cathode**), a copper(II) ion picks up two electrons and becomes a copper atom.

copper(II) ion + 2 electrons → copper atom

$$Cu^{2+} \quad + \quad 2e^- \quad \rightarrow \quad Cu$$

So copper transfers from the anode to the cathode. The impurities drop to the bottom of the beaker.

The photograph shows a factory where purification of copper is taking place. The man is walking over the copper electrodes.

Alloys

We don't use very many metals in their pure form. Metals are more useful to us when they are made into alloys. An alloy is a mixture of two elements where at least one is a metal.

The table gives information about some common alloys.

Alloy	Main metal in alloy	Use
amalgam	mercury	tooth fillings
brass	copper and zinc	hinges, screws, ornaments
solder	tin and lead	joining metals

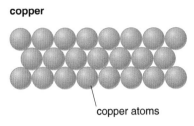

copper

copper atoms

The diagram on the right shows the arrangement of atoms in pure copper and in brass.

The properties of brass are different from the properties of the copper and zinc that make it up. In copper, the layers of copper atoms can slide over each other, In brass, the copper and zinc atoms are different sizes. The layers do not slide over each other easily. Brass is harder than copper or zinc. This makes it more useful. Brass can be used for door handles, but copper and zinc would not be strong enough.

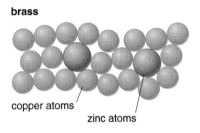

brass

copper atoms

zinc atoms

 Suggest what properties amalgam needs to have to be used as fillings.

Smart alloys

Nitinol is one of the names for a family of nickel-titanium alloys that are **smart alloys** or shape-memory alloys (SMA). The alloy was named after the Nickel Titanium Naval Ordnance Laboratory where it was discovered in 1961.

It is a mixture of approximately equal amounts of nickel and titanium. It can remember its original shape. A piece of nitinol wire can be bent into the shape of a paper clip. It works well as a paper clip. However, it goes back to being a straight piece of wire if an electric current passes through it or it is put into hot water.

The actual discovery of the shape memory property of Nitinol came about by accident. At a laboratory management meeting, a strip of Nitinol that was bent out of shape many times was presented. One of the people present heated it with his pipe lighter, and surprisingly, the strip stretched back to its original form.

Originally developed for military uses, smart alloys are now widely used including in a number of medical applications. Spectacle frames made of nitinol may go out of shape in use. Putting them in hot water will restore the original shape. Nitinol is also used to make hooks on wires that hold tendons to the bone in shoulder surgery.

Other smart alloys include:

- copper-aluminum-nickel alloys
- copper-zinc-aluminum alloys
- iron-manganese-silicon alloys.

keywords

alloys • anode • cathode • electrolysis • electrolyte • smart alloy

The search for copper

Copper is an extremely important metal and scientists are always looking for new supplies of it.

Divers recently recovered a 17 tonne boulder of nearly pure copper on the bed of Lake Superior in Canada. Similar boulders have been found in the past, but this is the largest.

Copper boulders are purified on the side of the lake by electrolysis. The diagram shows how this is done.

Today, copper purification usually takes place where there are relatively cheap supplies of electricity available. This is because very large amounts of electricity are required to make one tonne of copper. Often this purification is near hydroelectric power stations or where the manufacturer can negotiate an economical rate for electricity.

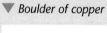
▼ Boulder of copper

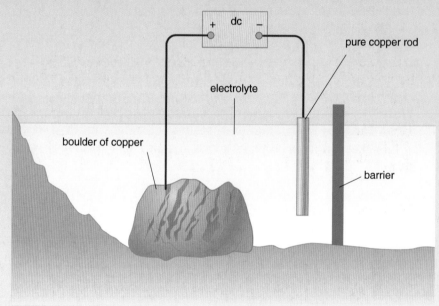

Questions

1 Suggest why it is possible for copper to remain on the bed of a lake without reacting with anything.

2 What is the advantage of purifying copper at the lakeside?

3 Use the diagram to explain how the copper is purified by electrolysis by the side of the lake.

4 Write ionic equations for the reactions that take place during the electrolysis.

5 Chemists have shown by analysis that the boulder from Lake Superior contains 80% copper. What is the maximum mass of pure copper that can be obtained from the boulder. Hint: 1 tonne = 1000 kg.

6 What are the incentives to recycle more copper?

Problem or resource?

In this item you will find out

- about rusting
- the advantages and disadvantages of building car bodies from aluminium or steel
- about recycling car materials

The manufacturers of some new cars recycle materials from old cars to help reduce costs and to avoid using up scarce resources. Cars are built from steel, copper, aluminium, glass, plastics and fibres.

a Suggest what properties of glass need to be used for car windscreens.

b Aluminium is light but strong. Why do you think is it good for making car bodies?

Old cars at breaker's yards can provide spare parts for other cars. Some manufacturers buy old engines or gear boxes and remanufacture them up to the original specifications. The rest of the car is crushed and much of it is recycled. The table shows the percentage of different materials currently recycled.

Material	Percentage (by weight) recycled
iron and steel	62
copper	50
aluminium	40
glass	60
polymers including fabrics	50

A modern car usually has a steel body which is much thinner than the steel used 30 years ago. This is possible because steel makers can make steel sheet with a very uniform thickness. Also, the car makers are able to use rust-proofing techniques which stop car bodies rusting as they used to. The benefit to the manufacturer is a reduction of costs of production and the benefits to the car-owner is better fuel economy because the car is lighter and a longer life for the car.

Recycling materials is good because it saves resources and cuts down on the disposal of waste. But the value of an old car for scrap barely covers the cost of collecting it. For this reason cars are sometimes seen abandoned. When a car is broken up, not all the materials in the car are re-used.

Makers of new cars in the future will have to design their cars so that more of the materials in the car can be recycled. When they sell a new car they will have to agree to take it back at the end of its life. By January 2007 they will have to plan to recycle 85% of the car and by 2015 at least 95%.

Rusting

The diagram shows an experiment to find what is needed for **rusting** (**corrosion**) of iron to take place.

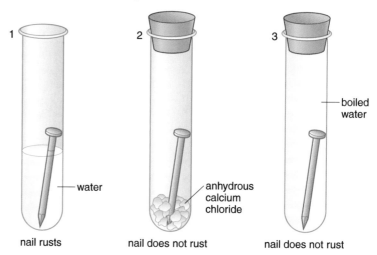

nail rusts   nail does not rust   nail does not rust

Test tube 1 is a control experiment. It shows that an iron nail rusts with air and water. Test tube 2 contains a nail inside a test tube with dry air. The anhydrous calcium chloride removes any water vapour. No rusting takes place here. Test tube 3 contains a nail in water but with no oxygen or air there. The nail does not rust. This experiment shows that rusting takes places when iron is in contact with water and oxygen (air). Rusting is speeded up by salt water or **acid rain**.

The rusting of iron is an **oxidation** reaction. Iron reacts with oxygen and water to form hydrated iron(II) oxide, $Fe(OH)_3.xH_2O$. When iron rusts the hydrated iron(III) oxide flakes off revealing a fresh surface for rusting. The rusting process can be shown by the word equation:

iron + oxygen + water → hydrated iron(III) oxide

Aluminium does not corrode even in damp conditions. This is because there is a thin coating of aluminium oxide on the surface of aluminium which does not flake off. This protects the surface from corrosion. Aluminium does corrode when it comes into contact with salt.

Car bodies

Alloys usually have different properties from the metals they are made from, which makes them more useful. Steel is an alloy containing iron with a very small percentage of carbon. Steel is harder and stronger than iron. It is also less likely to corrode than iron.

 The steel used today to make car bodies is much thinner than the steel used 30 years ago. Suggest advantages of this to both the car manufacturer and the car buyer.

Stainless steel is an alloy that does not rust at all. About 25 years ago a car was produced with a stainless steel body. This was more expensive than using normal steel but the car did not need painting. The photograph on the next page shows the stainless steel DeLorean sports car.

Some cars now have bodies made from aluminium rather than steel. This has several advantages. The same car made with an aluminium body rather than a steel body will be lighter because aluminium is less dense and so the owner will get better fuel economy. An aluminium car body will corrode less than a steel body and so the car will last longer.

However, an aluminium car body will be more expensive than the same car with a steel body. Also, repair costs are at least 25% more than for repairs of cars with steel bodies.

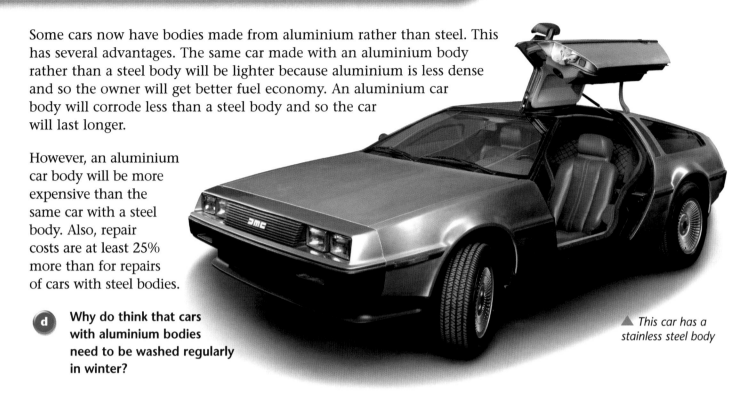

▲ This car has a stainless steel body

d Why do think that cars with aluminium bodies need to be washed regularly in winter?

Corrosion of two metals together

There are times when two metals might come in contact in a car. This can sometimes cause problems, for example, a car manufacturer is considering using metal rivets to fix panels of steel together. Either zinc or copper rivets could be used. Some experiments can be carried out to investigate what happens when the two metals are in contact. The following three experiments can be set up:

A one with an iron nail
B one with an iron nail in contact with a piece of zinc
C one with an iron nail in contact with a piece of copper.

Ferroxyl indicator can be used to show where rusting of iron is taking place. Areas that are pink are protected from corrosion. Areas that are green show where corrosion is taking place.

The diagram shows the results.

e What can you conclude from the results of the experiments with the three nails?

f Which rivets would you suggest are used? Explain your choice.

g The manufacturer also makes the car with an aluminum body. He is planning to bolt body panels together. Aluminium bolts are not strong enough. Would you suggest iron bolts? Explain your answer.

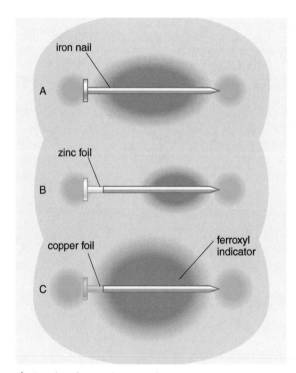

▲ Results of corrosion experiment

keywords
acid rain • corrosion • oxidation • rusting

Rusty cars

Thirty years ago cars rusted far more than they do today. There have been great advances in the treatment of steel to prevent rusting. Car manufacturers are able to give guarantees that the cars will not rust within 12 years.

Steel used for car making today is coated with zinc. This is called galvanising. The steel is usually dipped into a bath of molten zinc. This gives the steel protection against rusting.

Zinc is a reactive metal, which is more reactive than iron. It may seem strange that zinc should be able to protect steel from corrosion when it is more reactive and should need all the protection it can get. When oxygen in the air reacts with the surface of zinc, a very dense and impermeable coating of zinc oxide is formed. It is this physical barrier that protects the zinc surface from further attack.

If the surface of the galvanised steel becomes scratched, you would expect the exposed steel area to corrode. But the zinc and iron form an electrolytic cell. Rusting of iron involves losing electrons. In the electrolytic cell, zinc loses electrons more readily than iron. So the rusting of steel is reduced when zinc in is contact with the steel. If a metal lower in the reactivity series than iron was used to protect steel instead of zinc, corrosion would be speeded up.

Reactivity series of metals

Potassium	most reactive
Sodium	
Calcium	
Magnesium	
Aluminium	
Carbon	
Zinc	
Iron	
Tin	
Lead	
Hydrogen	
Copper	
Silver	
Gold	
Platinum	least reactive

(elements in italics, though non-metals, have been included for comparison)

▲ *Reactivity series*

Questions

1 What name is given to the protection of steel with zinc?

2 Zinc is more reactive than iron. Why does the zinc on the surface of the steel not corrode?

3 Write an ionic equation for the rusting of iron to produce iron(III) ions.

4 Which of the metals in the list, in contact with iron would speed up corrosion: aluminium, magnesium or lead?

5 A car manufacturer wants to use a metal washer between two lengths of steel exhaust pipe. Copper and zinc washers are available. Which washer should he use and why?

Air fit to breathe

In this item you will find out

- the composition of clean air

- how air pollutants are formed and their effects

- how atmospheric pollution is controlled

The photograph shows an aerial view of Santiago in Chile. This city has one of the biggest **atmospheric pollution** problems in the world.

The city is in a deep valley with high mountains around it. The direction of the wind does not usually blow the pollution out of the valley. It is difficult to see any distance because of atmospheric pollution.

So what causes atmospheric pollution? The table shows some of the common pollutants in the air, how they are formed and what effects they have.

Pollutant	How it forms	Effects
carbon monoxide	incomplete combustion in petrol or diesel engine	poisonous gas
oxides of nitrogen	formed in car engines from reaction of nitrogen and oxygen	acid rain and photochemical smog
sulfur dioxide	combustion of fuels containing sulfur, e.g. petrol	acid rain

Atmospheric pollution is caused by emissions from houses, vehicles and factories. As the population grows and industry develops the problem can get out of control. Motor cars and other vehicles are today regarded as the major cause of atmospheric pollution in towns and cities.

a Some countries are getting petrol companies to remove all sulfur from petrol. How will this reduce atmospheric pollution problems?

In Santiago, some cars are banned from using the roads on certain days. On one day all cars with registrations ending in 2 or 3 may be banned. On another day those ending in 7, 8 and 0 might be banned.

b Why does this reduce the atmospheric pollution levels?

The bar chart on the right shows the level of nitrogen oxides in the atmosphere in Santiago over the period of a week.

c Suggest what might have happened on Wednesday to change the level of nitrogen oxides in the atmosphere.

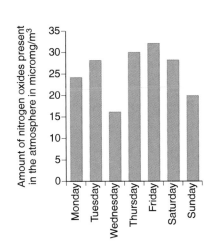

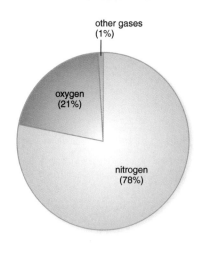

other gases (1%)

oxygen (21%)

nitrogen (78%)

What is in air?

We all take air for granted as it is always there. We know today that air is a mixture of gases. However, two hundred years ago many people believed that air was an element and not a mixture at all. Two scientists, Priestley and Lavoisier, did experiments to show that air was a mixture of gases. Their ideas were not accepted immediately.

d Why do you think new ideas in science are often rejected at first by other scientists?

The pie diagram on the left shows the percentage composition by volume of clean dry air.

e Which gas is in the atmosphere in the largest amounts? The table gives the typical composition of a sample of dry air.

Gas	nitrogen	oxygen	carbon dioxide	argon	other gases
Percentage	78.0	21.0	0.03	0.9	0.07

f Which is the reactive gas in the air?

g Which gas in the table is a compound?

Argon is one of a family of gases including helium, neon, krypton and xenon. These are called noble gases. They used to be called rare gases.

h Why do you think rare gas is a poor name for argon?

An electric light bulb contains a mixture of nitrogen and argon. A fluorescent light tube usually contains neon.

Because air is a mixture, its composition can vary from place to place. But the composition never varies much from that shown.

Carbon dioxide and oxygen balance

The percentage of carbon dioxide and oxygen in the air remains constant.

The diagram on the left shows a simple carbon cycle, which explains why the percentages remain constant. The processes of **combustion**, especially of carbon fuels, and **respiration** both use up oxygen and produce carbon dioxide.

$$C + O_2 \rightarrow CO_2$$
$$C_6H_{12}O_6 + 6O_2 \rightarrow 6CO_2 + 6H_2O$$

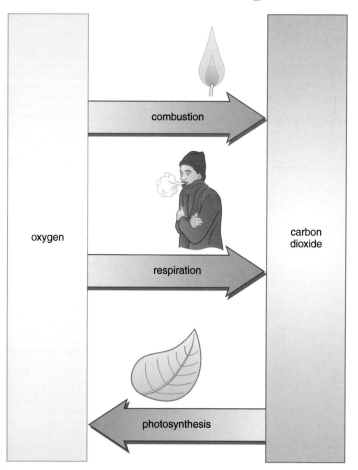

oxygen

combustion

respiration

photosynthesis

carbon dioxide

▲ Keeping oxygen and carbon dioxide in balance

Photosynthesis in green plants uses up carbon dioxide and produces oxygen.

$$6CO_2 + 6H_2O \rightarrow C_6H_{12}O_6 + 6O_2$$

There is a balance between respiration/combustion and photosynthesis that keeps the composition of the atmosphere constant.

The balance of carbon dioxide and oxygen in the air can be disturbed by human influences:

- **deforestation** – The burning of timber uses up oxygen and produces carbon dioxide.
- increasing energy consumption – The more fossil fuels that are burned the more carbon dioxide is produced.
- population increase – More oxygen is used in respiration and there is increased burning of fossil fuels.

The origin of the atmosphere

There have been many theories about how the Earth's atmosphere came to be. This is the most widely accepted one.

The Earth's original atmosphere, billions of years ago, probably came from gases that escaped from inside the Earth. These gases included carbon dioxide and steam, with smaller amounts of methane and ammonia. As the Earth cooled, the water vapour condensed to form the oceans. Some of the carbon dioxide dissolved in the water.

Then about 3000 million years ago, simple plant life in the sea converted carbon dioxide into oxygen by photosynthesis. Ammonia in the early atmosphere was converted into atmospheric nitrogen by bacteria. Nitrogen remains in the atmosphere because of its lack of reactivity.

Reducing atmospheric pollution

The World Health Organisation makes recommendations for air quality. In Lahore in Pakistan the levels of pollutants are 20 times greater than the WHO maximums. In a recent year, 6.4 million people in Pakistan were admitted to hospital with illnesses caused by air pollution. It is important to control atmospheric pollution because of the effect it has on humans and the environment.

A catalytic converter can be fitted to the car exhaust system to remove the pollutants carbon monoxide and nitrogen monoxide. It converts poisonous carbon monoxide and nitrogen monoxide in the exhaust gases into carbon dioxide and nitrogen.

carbon monoxide + nitrogen monoxide → carbon dioxide + nitrogen

$$2CO + 2NO \rightarrow N_2 + 2CO_2$$

ⓘ **Some people say that producing carbon dioxide is not acceptable. Suggest why.**

keywords

atmospheric pollution •
combustion • deforestation
• photosynthesis •
respiration

Photochemical smog

Photochemical smog is a type of air pollution. It forms when sunlight acts upon motor vehicle exhaust gases to form harmful substances such as ozone (O_3). Ozone causes breathing difficulties and can make respiratory problems worse. Photochemical smog can irritate the eyes, causing them to water and sting.

Motor vehicles produce exhaust gases containing oxides of nitrogen such as nitrogen dioxide (NO_2) and nitrogen monoxide (NO). When the nitrogen dioxide concentration is above clean air levels and there is plenty of sunlight, an oxygen atom splits off from the nitrogen dioxide molecule and reacts with oxygen molecules in the air to form ozone.

Nitrogen monoxide can remove ozone by reacting with ozone to form nitrogen dioxide and oxygen. When the ratio of NO_2 to NO is greater than three, the formation of ozone is the dominant reaction. If the ratio is less than 0.3, then the nitrogen monoxide reaction destroys the ozone at about the same rate as it is formed, keeping the ozone concentration below harmful levels.

Scientists measure the concentration of nitrogen oxides in the atmosphere and the hours of sunshine. Here are their results for one week.

	Monday	Tuesday	Wednesday	Thursday	Friday	Saturday	Sunday
NOx (ppb*)	80	80	160	90	80	90	50
Hours of sunshine	4	4	7	4	1	3	3
* ppb = parts per billion							

Questions

1 Nitrogen (N_2) and oxygen (O_2) combine in the car engine to form nitrogen monoxide. Write a symbol equation for this reaction.

2 Photochemical smog is more of a problem in Santiago or Pakistan than in the UK. Suggest why.

3 Nitrogen monoxide reacts with more oxygen to form nitrogen dioxide. Write a symbol equation for this reaction.

4 Write the symbol equation for the reaction of nitrogen monoxide and ozone.

5 Why do chemists monitor the NO_2:NO ratio?

6 Why do you think the level of photochemical smog and ozone is higher on Wednesday than Saturday? Refer to the data in the table.

7 The average value for the concentration of nitrogen oxides is 90 ppb. Why is this value alone not very useful when comparing the air pollution in different cities? What extra information should be given?

Closer and hotter

In this item you will find out

- that chemical reactions take place at different rates

- how temperature affects the rate of reaction

- how concentration affects the rate of reaction

Have you ever eaten chilli con carne? It is a famous dish made from beef, tomatoes and kidney beans. You can buy kidney beans in cans, but if you want to used dried beans then you have to soak them overnight and boil them in a saucepan in lots of water for about two hours.

If you have a pressure cooker, you can speed up the cooking of the dried beans. In a pressure cooker, the beans are cooked in about one-quarter of the time. The steam produced by the boiling water cannot escape and so the pressure inside the pressure cooker builds up. As the pressure builds up, the temperature of the boiling point of water increases above 120 °C. At this higher temperature the cooking process is speeded up.

This example shows that reactions can be speeded up by using a higher temperature or a higher pressure.

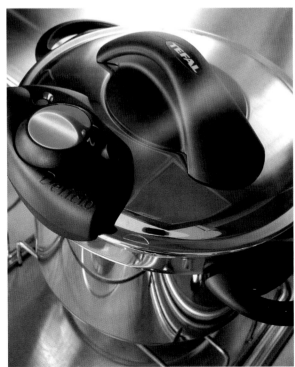

▶ Using a pressure cooker speeds up cooking times

Amazing fact

You may think that a pressure cooker is a new invention. The first pressure cooker was used over 500 years ago.

Increasing the temperature

A reaction takes place when particles of the **reactants** collide with each other. The more collisions that take place the faster will be the reaction. One way of speeding up a chemical reaction is by increasing the temperature. Increasing the temperature makes the particles move faster. This is shown in the diagram.

 More collisions as the temperature rises

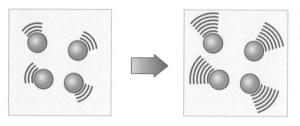

if the acid is heated the particles move faster

As a result there is a greater frequency of collisions (more collisions each second). When a collision occurs, a reaction only takes place when the colliding particles have more than a certain amount of energy (called the activation energy). At a higher temperature the particles possess more energy and so more collisions will have enough energy to undergo reaction. This results in a faster reaction.

a Choose the correct words to complete these sentences:
  (i) As temperature increases the average energy of the particles increases/decreases/stays the same.
  (ii) The number of effective collisions increases/decreases.
  (iii) This makes the reaction faster/slower.

Increasing the concentration

Another way of increasing the **rate of reaction** is to increase the **concentration** of one of the reactants. Increasing the concentration means there are more particles in a given volume. The particles are closer together. Again, more collisions will happen each second – the frequency of the collisions will be greater.

 More particles means more collisions

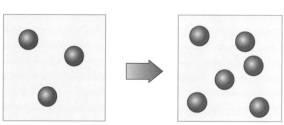

if the concentration of the acid is increased, there are more acid particles in the same volume of water

If the reactants are gases, for example hydrogen and oxygen, then increasing the pressure is the same as increasing the concentration. Both result in the particles being more crowded together which speeds up the reaction.

b Choose the correct words to complete these sentences:
  (i) As concentration increases the average energy of the particles increases/decreases/stays the same but the number of effective collisions increases/decreases.
  (ii) This makes the reaction faster/slower.

Investigating rate of reaction

The diagram on the right shows the apparatus that can be used to investigate the rate of a reaction. We are going to react marble chippings (calcium carbonate) and dilute hydrochloric acid together to produce calcium chloride, water and carbon dioxide.

calcium carbonate + hydrochloric acid → calcium chloride
+ water + carbon dioxide

$$CaCO_3 + 2HCl \rightarrow CaCl_2 + H_2O + CO_2$$

The solid line on the graph below shows the results.

The graph that we get is a curve. The graph starts steeply. As the reaction proceeds the gradient of the graph reduces and reduces. Finally when the reaction has stopped the graph is horizontal. The reaction stops when one of the reactants is used up.

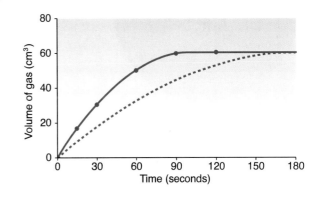

carbon dioxide

hydrochloric acid

marble chippings

 c After how long does the reaction stop?

We repeat the reaction. This time the hydrochloric acid is mixed with an equal volume of water. The concentration of hydrochloric acid is half what it was in the first experiment.

The results are shown by the dotted line on the graph. Now that the concentration is less, the reaction takes longer. It is a slower reaction. The graph is less steep and takes longer to become horizontal. The same volume of the gas carbon dioxide is produced because the same mass of calcium carbonate is used. The amount of **product** depends upon the quantity of reactant used.

 d In the first experiment 0.17 g of calcium carbonate was used. What would be the final volume of gas collected if 0.34 g was used?

The rate of the reaction at any point can be calculated by measuring the slope (gradient) of the curve at that point. The initial rate of the first reaction can be obtained using the following formula.

$$\text{Initial rate} = \frac{\text{volume of gas}}{\text{time}} = \frac{40}{30} = 1.3 \text{ cm}^3/\text{s}$$

As the reaction progresses the slope decreases.

 e What is the slope when the reaction has finished?

The table gives the time taken to complete a reaction at different temperatures.

Temperature (°C)	10	20	30	40	50
Time (s)	180	90	40	15	5

keywords

concentration • products • rate of reaction • reactants

 f People sometimes say that increasing the temperature of the reaction doubles the rate of reaction. Look at the table. Is this statement always true?

Fireworks

In the UK fireworks are graded. Party poppers are graded 1. This means they can be safely used indoors. Garden fireworks are graded 2 or 3 and large display fireworks are graded 4 or 5. The grading depends mainly on the power of the explosive mixture.

Party poppers are indoor fireworks. They contain small amounts of an explosive. The explosive is a mixture of a fuel, a chemical that supplies oxygen and other chemicals to give special effects.

When the string is pulled, the explosive is ignited. The explosion takes only a tiny fraction of a second. It produces a sound, some heat energy and a large volume of gaseous products. It is this that forces out the streamers.

Sparklers should be used outside because they burn at very high temperatures. They are a grade 2 firework. A sparkler mixture does not contain an explosive. The usual mixture for making the sparkler is iron filings, aluminium powder, barium nitrate and a glue to hold the materials together. A slurry of the ingredients is made and iron wire is dipped in and pulled out. This is repeated a number of times. Finally some primer paint is put on the end of the sparkler to make it easier to catch alight.

Once alight, the products include solid barium oxide, oxygen and nitrogen gases. The aluminium burns in the oxygen produced to form aluminium oxide, Al_2O_3. The gases cause the iron filings to be ejected. They catch alight in the high temperatures to produce the sparkle.

Garden and display fireworks are often bright colours. Small amounts of metal or metal compound give the firework the characteristic colour.

The table gives some colours produced by different metal or metal compound.

Metal or metal compound	Colour
lithium	red
sodium	orange/yellow
potassium	lilac
copper	blue/green
magnesium	white

Questions

1 There is a smell associated with a party popper but not with a sparkler. What do you think causes this smell?

2 A small amount of explosive that explodes in a confined space can cause serious problems. Why does this not cause a problem in a party popper?

3 Which chemical in the sparkler provides the oxygen needed?

4 Write a balanced symbol equation for the burning of aluminium in oxygen.

5 Write a balanced symbol equation for the burning of iron in oxygen to form the iron oxide Fe_3O_4.

6 Balance the equation for the reaction in the sparkler:
$Al + Ba(NO_3)_2 \rightarrow BaO + N_2 + Al_2O_3 + O_2$.

7 Why is unwise to let young children hold sparklers?

8 What colour would you expect a firework to be if the following are added to the explosive mixture?
   a sodium and magnesium
   b magnesium and copper

Explosions and catalysts

In this item you will find out

- how catalysts change the rate of chemical reactions

- how increasing the surface area of a reactant speeds up a reaction

- about explosions

An **explosion** is a very fast reaction that releases a large amount of products as gases.

We expect explosions to be caused by high explosives such as trinitrotoluene (TNT) or dynamite. Chemicals such as ammonium nitrate can also be explosive. On 21 September 2001 a huge explosion occurred in a fertiliser factory in Toulouse in France involving 200–300 tonnes of stored ammonium nitrate. As a result 31 people died and the explosion caused damage over a wide area.

However, serious explosions can happen in factories where flour, custard powder or sulfur are used. The photograph shows the results of an explosion at Blaye in France in 1997.

The explosion at Blaye was caused when dust from the grain used to make flour formed an explosive mixture with air. In this accident 11 people were killed.

There are important guidelines to be followed in factories and storage facilities that handle powders.

1   Workers must wear special boots which do not cause friction against the floors.

2   There must be no flames, no smoking and no hot surfaces.

▲ *Explosion at Blaye*

3   Factories must have efficient ventilation systems which remove the air and replace it with fresh filtered air.

a Which guidelines are intended to stop the ignition of explosive mixtures?

b Which guidelines are intended to stop a build up of explosive mixtures in the factory?

Amazing fact

In one week in 2005 over 100 people were killed in China in explosions in coal mines involving explosive mixtures of coal dust and air.

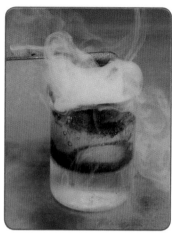

Catalysts

Hydrogen peroxide is widely used as a bleach. Some people use it to bleach hair.

Like water, it is compound of hydrogen and oxygen. It has a formula H_2O_2.

It decomposes very, very slowly to form water and oxygen. The photographs on the left show some hydrogen peroxide in a beaker before and after adding powdered manganese(IV) oxide.

The manganese(IV) oxide acts as a **catalyst**. A catalyst speeds up chemical reactions.

Examiner's tip

Candidates often think that all catalysts are enzymes. Enzymes are only one type of catalyst.

The manganese(IV) oxide speeds up the reaction of hydrogen peroxide. A catalyst is not used up and remains unchanged at the end of a reaction. Only a small amount of catalyst is needed to speed up the reaction of a large amount of reactants. A catalyst is also specific to a particular reaction. A catalyst for one reaction may not catalyse a different reaction. Catalysts do not alter the amount of product being made in a reaction, they just allow the reaction to happen more quickly.

c What is the benefit to a manufacturer of producing the same amount of product more quickly?

The equation for the hydrogen peroxide reaction is:

$$\text{hydrogen peroxide} \rightarrow \text{water} + \text{oxygen}$$
$$2H_2O_2 \rightarrow 2H_2O + O_2$$

The graph on the left shows the decomposition of $25\,\text{cm}^3$ of hydrogen peroxide with $0.2\,\text{g}$ of manganese(IV) oxide.

d Read the volume of gas collected after 30 s from the graph.

e What is the total volume of gas collected when all the hydrogen peroxide has reacted?

f How can you tell from the graph that the reaction is faster after 30 s than it is after 60 s?

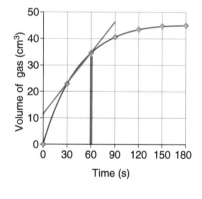

Industrial catalysts

The table gives some examples of catalysts used in industrial processes.

Catalyst	Reaction
titanium(IV) chloride	polymerisation of ethene
vanadium(V) oxide	contact process to make sulfuric acid
iron	Haber process to make ammonia
nickel/rhodium alloy	hardening vegetable oils to make margarine
platinum	making nitric acid from ammonia

▲ Platinum gauze used as a catalyst in making nitric acid

Increasing surface area

In the last item we looked the reaction between marble and hydrochloric acid (page 137). We are going to repeat the experiment but this time with the same mass of powdered calcium carbonate.

The line graph on the left shows the results. The graph line for the original experiment is there for comparison.

(page 137)

 Is this reaction faster than the original experiment? How can you tell from the graph?

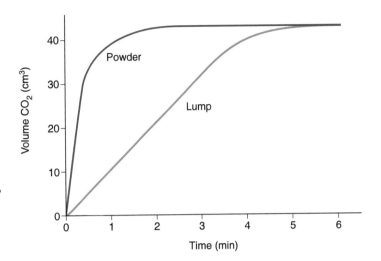

The photographs below shows the difference between a lump reacting and a powder reacting.

The reaction is faster because the powder has a larger **surface area** for the acid particles to collide with. There are more collisions per second (a greater frequency of collisions) so the reaction happens faster. This is shown in the diagram below.

If the limestone is crushed, the surface area is bigger because more surface is exposed

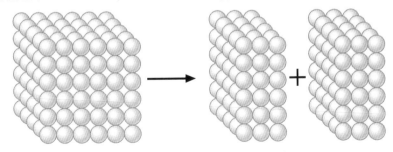

▲ *A larger surface area means more collisions. In each case, the acid particles collide with the limestone more frequently, and so the reaction will get faster*

keywords
catalyst • explosion • surface area

h **Catalysts are often in the form of fine meshes or pellets. Suggest why this is better than large solid lumps.**

Making tablets dissolve

Amanda works for a pharmaceuticals company that makes painkillers that can be bought over the counter. One particular painkiller can be bought as a round pill, as a lozenge-shaped pill with a sugar coating or as a powder.

These painkillers work by dissolving in the stomach. The painkilling active ingredients are then able to attack the source of the pain.

Some people prefer to use pills because they are handy to carry around and they are easier to swallow. The powders have to be mixed with water and drunk and they usually taste horrible.

Amanda carries out an experiment to find out how quickly each painkiller dissolves.

She test three different products A, B and C. She drops a single dose of each painkiller into acid similar in concentration to the acid in the stomach. She uses the same volume of acid each time.

Her results are shown on the graph. Three lines A, B and C are plotted.

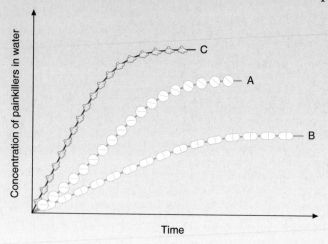

Questions

1 Why do you think it is an advantage for painkillers to dissolve quickly?

2 Which type of painkiller has the fastest rate of reaction?

3 Explain why this painkiller dissolves the quickest.

4 Suggest a reason why the sugar-coated lozenge dissolves slowly.

5 It is important to take the correct dose of painkiller. Suggest why pills are better than powders for doing this.

6 In the photograph there are some capsules containing very tiny pellets. When these are swallowed the outside of the capsule quickly breaks down and the tiny pellets escape. Explain why do you think these are better than a tablet?

7 What evidence is there in the graph that the different products contain different amounts of painkiller?

C2a

1 Years ago gloss paints had to be stirred before use and from time to time during use. It is not necessary today. Suggest why. [2]

2 For many centuries cloths have been dyed using plant dyes such as woad. Now synthetic dyes are used.

Suggest two reasons why synthetic dyes have replaced natural dyes. [2]

3 A wood varnish consists of a natural resin dissolved in an organic solvent.

Describe the change that takes place when the varnish dries. [1]

4 Describe two things that happen when a gloss paint dries. [2]

5 The table gives the percentage composition of two oil-based paints.

Paint	Binding medium	Solvent	Pigment
undercoat	40	20	20
top coat	60	20	10

Undercoat must provide a good base colour, covering up any paint underneath. Top coat must give a tough, shiny, final finish.

Describe how the composition of these paints makes them suitable for their purpose. [4]

C2b

1 Write down the name of a construction material made from:

a bauxite (aluminum ore) [1]
b clay [1]
c sand. [1]

2 Put these three rocks in order of increasing hardness:

granite   limestone   marble [2]

3 Which two materials heated together produce cement? [2]

4 Use chemicals from the list to write the word equation for the thermal decomposition of limestone.

**calcium carbonate   calcium oxide
carbon dioxide   water** [2]

5 Explain why steel-reinforced concrete can be used for weight-bearing beams in construction. [3]

6 Write a balanced symbol equation for the thermal decomposition of calcium carbonate, $CaCO_3$. [2]

7 Granite, limestone and marble are three rocks used in the construction industry.

a Which rock is produced by the crystallization of molten rock from the magna? [1]
b Which rock is produced by the action of high temperatures and pressures on existing rocks? [1]
c Which rock is sedimentary? [1]
d Which rock cannot contain fossils? [1]

C2c

1 a What causes an earthquake? [1]
b Where on the Earth are earthquakes most likely to happen? [1]

2 Explain why some igneous rocks are made up of small crystals and others of large crystals. [2]

3 The theory of plate tectonics is a fairly recent one. Suggest difficulties scientists have had discovering the truth about the Earth's crust. [3]

C2d

1 Brass is an alloy used for making screws for fixing wood.

a Which metals are used to make brass? [2]
b Suggest two properties of brass that make it suitable for this use. [2]

2 a Describe some of the problems of recycling copper. [2]
b Why is it particularly important to recycle copper? [2]

3 Duralumin is an alloy made from aluminium and copper. It is denser than pure aluminium but stronger than pure aluminium or pure copper.

a Despite being denser than pure aluminium, duralumin is used for building aircraft rather than pure aluminium. Why is this? [1]
b Overhead power cables are made from pure aluminium rather than duralumin. Suggest why. [1]
c Pure aluminium costs £800 and pure copper costs £1200 per tonne. Work out the cost of the metals used to produce 1 tonne of duralumin containing 10% copper. [2]

4 Silver can be purified in a similar way to copper. The electrolyte is silver nitrate solution, $AgNO_3$.

  a Write down the symbols for the ions in silver nitrate solution. [2]

  b Write the ionic equation for the change at the positive electrode. [2]

  c Write the ionic equation for the change at the negative electrode. [2]

C2e

1 Explain why steel ship hulls have to be examined frequently for rusting. [2]

2 Suggest reasons why steel is used instead of iron for car bodies. [3]

3 Write the word equation for the rusting of iron. [2]

4 Explain the advantages and disadvantages of using aluminium rather than steel for car bodies. [4]

C2f

1 Look at the graph showing the number of deaths each day in London between 1st December and 15th December 1952. It also shows the concentrations of smoke and sulfur dioxide.

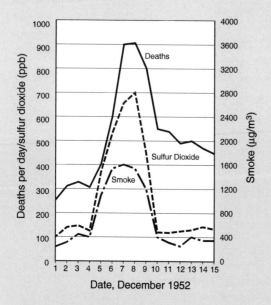

a How many deaths occurred on 5th December 1952? [1]

b What is the relationship between the number of deaths and the concentration of sulfur dioxide? [2]

c Is there a similar correlation between the number of deaths and the concentration of smoke? Explain your answer. [2]

2 The original atmosphere of the Earth included hydrogen and helium. Why did they escape from the atmosphere? [1]

3 **a** How are nitrogen oxides formed in a car engine? [1]

  b Why does this reaction not take place when a splint is burning in ordinary air? [1]

4 The diagram summarises an experiment using iron filings.

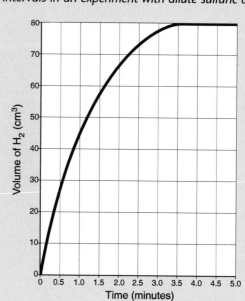

Explain the results of the experiment. [2]

5 Write a balanced equation for the reaction that takes place in a catalytic converter to remove carbon monoxide and nitrogen monoxide. [3]

6 The percentage of nitrogen has remained constant for a long time. Why is this? [3]

C2g

1 The graph shows the volume of hydrogen collected at intervals in an experiment with dilute sulfuric acid.

a After how long is the reaction complete? [1]
b After how long are half the reactants used up? [1]
c Calculate the average volume of gas produced each second in the first minute. [2]
d Why is your answer to **c** not a useful measure of the rate of reaction? [1]

2 Refer to question **1**.

Draw a tangent to the curve at one minute and calculate the rate of reaction at this time. [4]

3 Which one of the following graphs would be most useful in establishing that the rate of reaction is directly proportional to concentration (c) of the reactant? (T represents time.)

A c against T
B 1/c against 1/T
C c against 1/ T
D c against c × T [1]

C2h

1 The mass of a catalyst was determined at intervals during a reaction. Which one of the graphs A–D would be obtained? [1]

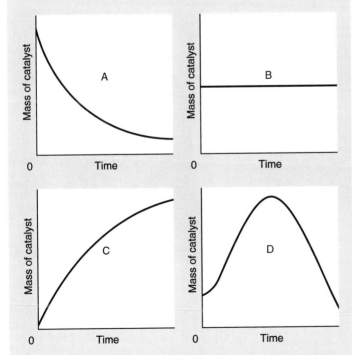

2 An experiment was carried out using marble chips and dilute hydrochloric acid to investigate the effects of particle size on the rate of reaction.

$$CaCO_3 + 2HCl \rightarrow CaCl_2 + CO_2 + H_2O$$

A large marble chip (mass 0.4 g) was placed in a conical flask and the flask placed on a top pan balance. 25 cm³ of hydrochloric acid was added to the flask and a plug of cotton wool was placed in the neck of the flask. The reading on the balance was noted at intervals.

The results are shown in the table.

Time (min)	0	2	4	6	8	10	11	12
Total loss in mass (g)	0.0	2.2	2.9	3.3	3.6	3.7	3.7	3.7

a Plot a graph of the total loss of mass (on the y-axis) against time. [3]
b When was the reaction fastest? [1]
c After how many minutes was the reaction completed? [1]

The reaction was repeated using 0.4 g of powdered marble and a fresh 25 cm³ sample of dilute hydrochloric acid.

d On the same graph, sketch the graph that would be obtained with powdered marble. [3]
e What can be concluded from these two experiments? [1]
f Explain your conclusion in **e**. Use ideas about particles in your answer. [2]

3 In an experiment to compare different ions as a catalyst for a certain reaction, the following results were obtained.

	Temperature (°C)	Substance tested as a catalyst	Time for reaction to be completed (s)
A	20	cobalt(II) chloride	18
B	20	sodium nitrate	36
C	20	cobalt(II) nitrate	12
D	30	cobalt(II) nitrate	8
E	20	sodium chloride	40

a Why should D not be used in any comparison? [1]
b Which substance gives the greatest increase in the rate of reaction? [1]
c Which substance is least effective as catalyst? [1]
d Which ion is most effective as a catalyst? [1]

P1 Energy for the home

I am worried about global warming. How do I change my lifestyl to minimise its effects?

Is my mobile phone going to give me cancer? Is it safe for me to cook my food by microwaves?

Why should I bother about the efficiency of my house heating system?

- Energy for heating our homes not only costs money, but its production also has a major affect on our environment. Knowing how to use that energy sensibly is vital for our future.

- Most of the energy which comes from the Sun arrives at the Earth as infrared and light waves and this can be used for heating houses. Infrared and light are part of the electromagnetic spectrum, a range of waves which we use for global communications – the Internet relies on them.

- But there are other types of wave as well. Waves from earthquakes create huge damage and loss of life, but also provide vital evidence about the internal structure of our planet.

What you need to know

- Energy can be transferred from one form to another.
- Heat energy passes readily through conductors and slowly through insulators.
- Solids, liquids and gases are made of particles.
- Light is a wave that travels in straight lines and can be reflected.

Warming up

In this item you will find out

- about energy flow
- about specific heat capacity
- about specific latent heat

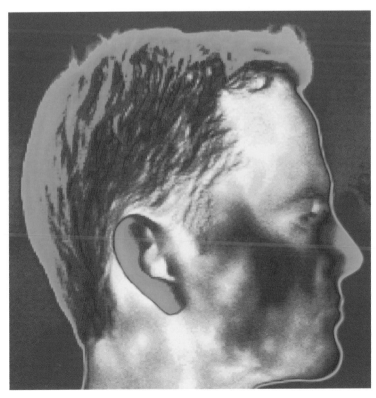

Tea made from cold tap water is disgusting. The **temperature** of the water is only about 10°C, not enough to make the tea properly. To get the water to 100°C, you need to add heat to it. So you put it in a kettle.

What's the quickest way of making tea? Easy – heat just enough water to fill a mug.

Suppose you put too much water in the kettle? It takes longer to boil, but why? The water needs more energy and the kettle can only supply so much.

Most electric kettles switch themselves off when the water starts to boil. Will your tea be hotter if the water boils for longer?

Surprisingly, it won't. Water boils at 100°C, no matter how much heat you add to it!

Heat and temperature are not the same things. Temperature is a measurement of how hot or cold an object is on a chosen scale while heat is a measurement of energy on an absolute scale.

Temperature is measured in °C and **heat energy** is measured in joules (J).

Temperature can be shown by the range of colours in a **thermogram**.

You can see from the photograph of the head that the cheeks are red and the nose is blue. This is because the cheeks are hot and the nose is cold.

a Describe the difference between heat and temperature.

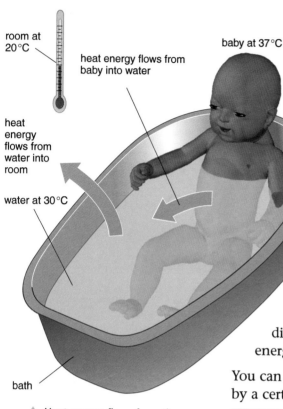

room at 20°C

heat energy flows from baby into water

baby at 37°C

heat energy flows from water into room

water at 30°C

bath

▲ Heat energy flows from the baby into the bath water

Energy flow

Heat energy always flows from a hot object to a cooler one. This causes hot objects to cool down and cool objects to warm up. If the water in a baby's bath is only 30°C, the baby will get cold because the heat energy will flow from the baby into the water.

b Suppose the bath water is at 39°C? Which way will the heat energy flow?

c Explain why the bath water eventually ends up at room temperature.

Specific heat capacity

The **specific heat capacity** (shc) of a material tells you how much energy, in joules, you need to raise the temperature of 1 kg of the material by 1°C. Specific heat capacity is measured in units of joules per kilogram per degree Celsius (J/kg/°C). It is different for different materials because some materials need more energy than others for the same temperature change.

You can calculate the energy needed to raise the temperature of a material by a certain amount by using this equation:

$$E = mcT$$

where E = heat energy added (J)
m = mass (kg)
c = specific heat capacity (J/kg/°C)
T = temperature change (°C)

For example, how much heat energy would you need to raise the temperature of 3 kg of glass from 12 to 20°C if the specific heat capacity of the glass is 700 J/kg/°C?

$E = mcT$
$E = 3\,\text{kg} \times 700\,\text{J/kg/°C} \times 8\,\text{°C} = 16{,}800\,\text{J}$
$E = 16.8\,\text{kJ}$

d Show that you need 50 kJ of heat energy to raise the temperature of 5 kg of steel from 10 to 30°C if the specific heat capacity for steel is 500 J/kg/°C.

From ice to steam

If you want to make steam from a solid block of ice at 0 °C then you need to add a lot of energy.

You need to add energy to the ice to melt it – to make it into a liquid at 0°C. Each kilogram of ice needs 330,000 J for this. The energy breaks bonds between the water molecules, giving them freedom to move past each other and become a liquid. While the ice is melting, energy is being added, but the temperature of the water does not rise.

Then you need more energy to increase the temperature of the water to 100°C, another 420,000 J for each kilogram. This energy increases the kinetic energy of the molecules.

Finally, you need another 2,300,000 J to boil 1 kg of water into steam giving the molecules enough energy to break completely free from each other and become a gas. While the water is boiling, energy is being added, but the temperature of the water does not get hotter than 100 °C.

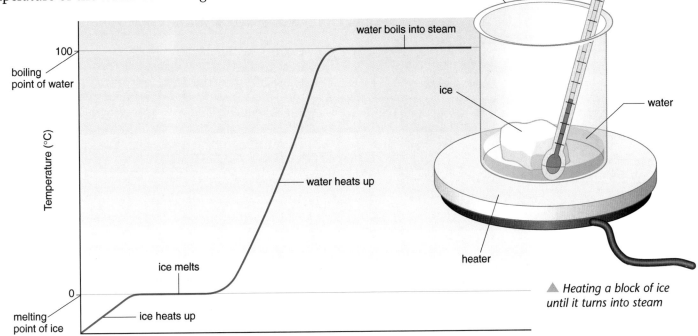

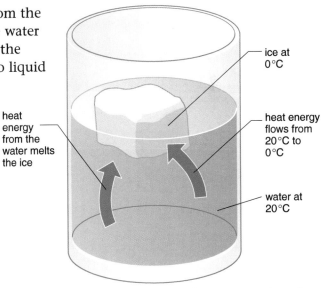

▲ Heating a block of ice until it turns into steam

Specific latent heat

When you add ice to a glass of water, heat energy flows from the water at 20 °C to the ice at 0 °C and the temperature of the water drops. This energy is needed to break the bonds that hold the water molecules in place in the solid ice and change it into liquid water. It is called latent heat.

The **specific latent heat** of a material tells you how much energy is needed to boil or melt 1 kg of that material. It is different for different materials. If you want to melt a 1 kg block of ice then you need to add 330,000 J of energy. So the specific latent heat of ice is 330,000 J/kg.

The energy needed to change the **state** of a solid, liquid, or gas is calculated with this formula.

$$E = ml$$

where E = heat energy added (J)
m = mass (kg)
l = specific latent heat (J/kg)

So how much energy is needed to melt a single 0.02 kg ice cube?

$E = ml$
$E = 0.02 \text{ kg} \times 330,000 \text{ J/kg} = 6,600 \text{ J}$
$E = 6.6 \text{ kJ}$

▲ Heat energy from the water melts the ice

 Show that you need 165,000 J to melt 0.5 kg of ice.

Cheap heat

Janet and Allan use a heat pump to keep their home warm in winter. The pump makes use of latent heat to move heat energy from the cold ground outside into the warm house.

The key ingredient is the refrigerant, a liquid which boils at about room temperature.

The pipes under the ground outside are kept at a low pressure, forcing the liquid to turn into a gas. As it boils, the liquid takes in latent heat from the ground around the pipes.

The gas then passes through a compressor on its way into the house. The compressor increases the pressure of the gas so that it condenses back into a liquid. This happens as the gas passes through pipes in the floor of the house, releasing latent heat into the house.

After passing through an expansion valve, the liquid passes underground once more and the cycle starts again. The house gets warmer as their garden gets colder!

Janet and Allan can also use their heat pump to keep their house cool in summer. All they have to do is reverse the direction of flow of the refrigerant through the pipes.

The pipes in the house are now at a low pressure forcing the refrigerant to boil, taking in heat energy. The gas condenses again in the high pressure conditions of the pipes under the ground, releasing heat energy. The inside cools down and the outside gets warmer.

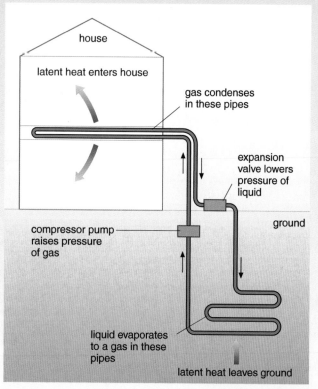

house

latent heat enters house

gas condenses in these pipes

expansion valve lowers pressure of liquid

ground

compressor pump raises pressure of gas

liquid evaporates to a gas in these pipes

latent heat leaves ground

Questions

1   Butane boils at 0 °C and is therefore suitable as a heat pump refrigerant. Explain why water is not suitable.

2   Describe what happens to the particles in a liquid as it turns into a gas. Explain why this requires the addition of latent heat.

3   Describe, in detail, how a heat pump warms a house in winter.

4   Although heat pumps have been around for a long time, very few houses in the UK use them. Suggest reasons why.

Keeping heat in

In this item you will find out

- about conduction, convection and radiation

- how to calculate the cost of different energy-saving methods

- how to calculate energy efficiency

In the UK, the average outside temperature is about 5 °C in winter. So we usually stay indoors where can heat our houses to 20 °C.

Look at the thermogram of the house. It shows the infrared radiation given off by the surface of the house. Each colour shows a different temperature – red for hot, blue for cold.

a Which part of the house is losing lots of heat?

b If the heat energy that escapes from the house is not replaced, what will happen to the temperature of the house?

The roof is blue. It is well insulated, with loft insulation to slow down the flow of heat energy. This cuts down on the cost of heating the house. But insulating the loft space is expensive. Can it be paid for by the money saved on the heating bills?

We can heat our houses in different ways and we can work out which methods are most efficient. We can also keep our houses warm by different energy-saving methods and we can calculate whether these methods save us money in the short- or long-term.

Any use of energy from fuels to heat our houses has an environmental cost. The fuels that we use may be non-renewable, so waste of heat energy through our walls may be depriving future generations of those fuels.

There are many good reasons, other than immediate cost, why our homes should be well insulated.

Amazing fact

It costs about £10 billion to heat all houses in the UK.

Holding in heat

There are three ways in which heat energy is lost from a house:

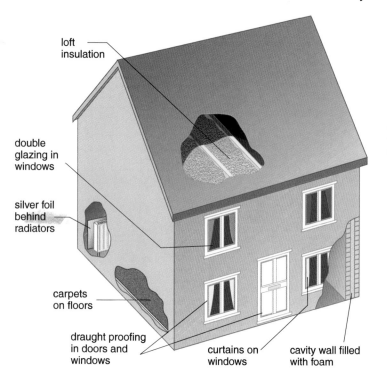

loft insulation

double glazing in windows

silver foil behind radiators

carpets on floors

draught proofing in doors and windows

curtains on windows

cavity wall filled with foam

- It is lost through solid materials, such as floors and walls, by **conduction**.
- It is lost from outside surfaces, such as the walls and windows, by **radiation**.
- It is carried away by the movement of air, such as draughts or the wind, by **convection**.

Energy saving strategies can be used to reduce heat energy loss:

- Fluffy insulation can be added, which exploits the fact that air is a bad conductor.
- Outside walls can be painted white, as white is a poor radiator.
- Draughts can be stopped to reduce loss of heat by convection.

c Look at the diagram on the left. Describe six different ways of insulating a house. What type of heat transfer is each insulation method stopping?

d Suggest ways in which you could reduce heat energy loss from everyday appliances.

Replacing heat

How do you keep a house at a steady comfortable temperature? You have to replace every joule of heat energy which escapes! There are several ways of doing this, each with their own installation and running costs.

Method of heating	Installation cost	Running cost
portable electrical heaters	cheap	expensive
electrical storage heaters	moderate	cheap
gas or oil central heating	expensive	cheap
wood or coal fires	expensive	expensive

There are other factors to consider when selecting a method of heating:

- How easy is it to install?
- How convenient is the fuel?
- How much pollution does it create?

So, portable electric heaters are easy to install. You just plug them into a socket. On the other hand, electricity is expensive. In addition, its production usually has pollution costs, such as global warming.

e State one good feature and one bad feature for each method of heating.

Counting the cost

Here are some facts about insulating a typical house.

Method of insulation	Installation cost (£)	Fuel saving (£ per year)	Payback time (year)
radiator foil	5	10	0.5
loft insulation	200	100	2
cavity wall insulation	900	150	6
double glazing	2,000	50	40

Putting a porch on the front of a house will save energy. But will it save money?

- The porch will cost about £1,000 to build.
- It will save about £50 per year in heating costs.

The **payback time** tells you how cost effective the porch will be.

$$\text{payback time} = \frac{\text{installation cost}}{\text{fuel saving}}$$

$$\text{payback time} = \frac{£1,000}{£50} \text{ per year} = 20 \text{ years}$$

This is better than double glazing, but worse than cavity wall insulation!

porch stops cold air entering house when people enter or leave

 Draught excluders on all doors and windows of a house costs £80 and save £20 of heating costs per year. Calculate the payback time.

Energy efficiency

The **efficiency** of a heating system tells you how much of the energy you put into the system goes into the house as useful heat energy. What doesn't become heat energy may be wasted as light or sound. Here are some examples.

keywords

conduction • convection • efficiency • payback time • radiation • total energy input • useful energy output

Heating system	Total energy input (J)	Useful energy output (J)	Efficiency
electrical heater	100	100	1.0
gas central heating	100	80	0.8
wood fire	100	40	0.4

Efficiency can be calculated using this equation:

$$\text{efficiency} = \frac{\text{useful energy output}}{\text{total energy input}}$$

g For every 100 J of energy put into a fan heater, only 95 J comes out as heat energy. What is its efficiency?

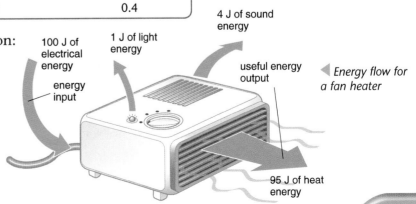

100 J of electrical energy

energy input

1 J of light energy

4 J of sound energy

useful energy output

Energy flow for a fan heater

95 J of heat energy

The efficiency expert

Jim works for EnergyCom. It sells electricity to households in the UK. His job is to advise people on ways of heating their houses.

Although EnergyCom makes its money by selling electricity, it doesn't want to invest in building new power stations. It makes more money by persuading people to make better use of its existing power stations.

EnergyCom uses coal-fired and nuclear power stations to make electricity, neither of which can be turned on or off suddenly. The most economic way of running them is to keep their output steady around the clock.

Electricity can't be stored, so it has to be used as soon as it is made. Jim's real job is to persuade people to buy cheap off-peak electricity. This is electricity generated after midnight, when many people are asleep and don't need electricity.

Jim persuades people to heat their houses with electrical storage heaters. They heat up blocks of concrete during the night, when electricity is cheap. Fans blow air through the blocks during the day to transfer the heat energy into the house.

'The real problem is the size of the heater,' says Jim. 'All the insulation around the blocks makes the heater quite large.'

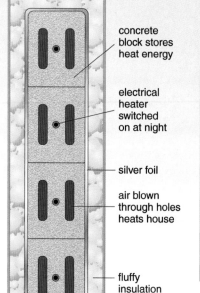

concrete block stores heat energy

electrical heater switched on at night

silver foil

air blown through holes heats house

fluffy insulation

▲ *How a storage heater works*

Questions

1 Explain why EnergyCom sells its electricity cheaply after midnight.

2 Explain how a storage heater works.

3 A small storage heater has 50 kg of concrete which is heated from 20 °C to 200 °C each night. The specific heat capacity of concrete is 800 J/kg/°C. How much heat energy does the heater store?

4 A household replaces all their electrical heaters with storage ones. Explain how they could, after a year, estimate the payback time for the new heaters.

5 Explain why the blocks need to be insulated.

How energy gets around

In this item you will find out

- about conduction, convection and radiation

- why many insulators contain trapped air

Have you ever wondered why most of your hair is on your head? It's to keep heat energy in your skull! A lot of heat energy is transferred from chemical energy in your brain. A head of hair allows a lot of that heat energy to be carried somewhere useful by your blood, instead of being lost to the surroundings.

Your body uses other mechanisms to keep you at a steady temperature of 37 °C.

- When you get too hot, extra heat energy radiates from your red skin.
- Layers of fat under your skin slow down the conduction of heat energy away from your body.
- The blood circulating around your body convects heat energy, sharing it evenly.

You can reduce heat energy flow from your body in different ways.

- You can wear clothes to stop hot air escaping from your skin.
- You can live in warm buildings to reduce the temperature difference between your body and its surroundings.

▲ A hat also keeps heat energy in your head

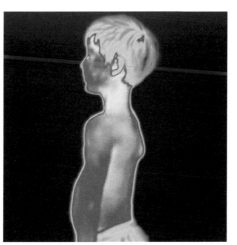

◄ This thermograph shows that heat is lost from your cheeks but held by your hair

Amazing fact

Your brain transfers more chemical energy to heat energy than any other organ in your body.

155

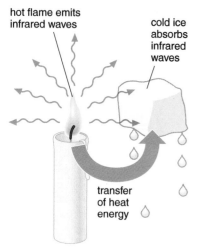

hot flame emits infrared waves

cold ice absorbs infrared waves

transfer of heat energy

▲ *Infrared radiation emitted by the hot flame is absorbed by the cold ice*

Infrared radiation

Everything loses heat energy through waves called **infrared** radiation. This is a wave, similar to light, but with a different wavelength so that it is invisible. You can feel it, but you can't see it.

Hot objects emit much more infrared radiation than cold ones. So a hot object placed in a cool place radiates heat energy until they both get to the same temperature.

Two properties of infrared radiation are used to control the heat energy flow from houses:

- White or shiny surfaces don't absorb infrared, they reflect it.
- White surfaces emit a lot less infrared than coloured ones.

a Why does a layer of shiny foil between a heater and the wall save money?

b Why can you save money by painting the outside walls of your house white?

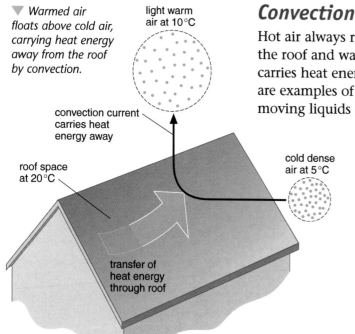

▼ *Warmed air floats above cold air, carrying heat energy away from the roof by convection.*

light warm air at 10°C

convection current carries heat energy away

cold dense air at 5°C

roof space at 20°C

transfer of heat energy through roof

Convection

Hot air always rises above cooler air, carrying heat energy away from the roof and walls. Hot water flowing in the central heating pipes carries heat energy from the heater to the radiators. Both of these are examples of convection, where heat energy is carried around by moving liquids or gases.

Natural convection happens because:

- A low density fluid always floats on top of a high density one.
- When a fluid is heated, it expands and so its density falls.

So when a gas or liquid is heated, it naturally rises up to float on top of the fluid which is at a lower temperature. Forced convection uses a pump to move the hot fluid from one place to another.

c How does wind remove heat from a house by forced convection?

d Suggest why planting trees around a house keeps it warm.

Conduction

Heat energy can flow through solid walls. How is this possible?

The particles which make up solids, liquids and gases never stop moving. The kinetic energy of this endless random motion is what we call heat energy. In solids, where particles are held together by forces between them, their heat energy makes the particles vibrate.

e What is the difference between conduction and convection?

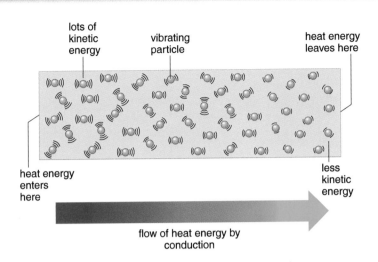

lots of kinetic energy

vibrating particle

heat energy leaves here

heat energy enters here

less kinetic energy

flow of heat energy by conduction

The vibrations of the particles carry heat energy by conduction from the hot end to the cold end

Conductors and insulators

Some materials conduct heat better than others.

• Conductors, such as metals, let heat energy flow through easily.
• Insulators, such as wood, only allow heat energy to flow through slowly.

Liquids, solids and gases can all let heat energy flow through them. Gases are insulators because their particles are far apart, making it difficult for them to transfer heat energy from one to another.

You can tell an insulator from a conductor by just feeling it. Conductors at room temperature allow heat energy to flow away from your skin, lowering its temperature. So, conductors feel cold to the touch. But insulators at room temperature do not take heat energy from your skin, so it stays at 37 °C.

Cavity wall insulation

Insulating materials in houses are full of tiny holes. This slows down heat flow through them in two ways:

• Heat energy has to be conducted through the thin solid between the holes.
• The air in the holes is trapped, so it can't use convection to take the heat energy away.

Cavity walls also slow down heat energy flow. Although convection currents in the air can carry heat energy from one side of the cavity to the other, the narrowness of the gap forces the two streams of air close to each other, slowing them down. Of course, filling the cavity with foam or wool which is full of tiny holes stops the flow of air completely and provides far better insulation.

 How do insulating materials slow down the transfer of heat energy?

Amazing fact

Wrapping accident victims in silver foil helps to keep them warm.

Examiner's tip

Know the difference between conduction, convection and radiation.

▼ *Cavity wall insulation*

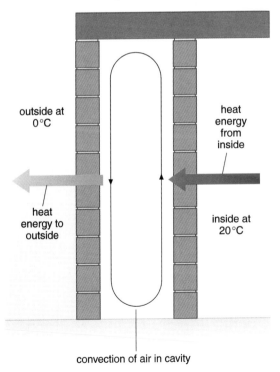

outside at 0 °C

heat energy from inside

heat energy to outside

inside at 20 °C

convection of air in cavity

keywords

cavity walls • infrared

Mission to Mars

Within a few years, a team of astronauts will probably be sent to Mars. The journey will take many months and there will be many dangers, so the designers of the spacecraft need to make the spacecraft as safe as they can. One of the things they must consider is the temperature inside the spacecraft.

Energy from the Sun keeps our planet at a comfortable temperature. The Sun will also deliver energy to the spacecraft on its voyage, but as it moves away from the Sun, that amount of energy will decrease.

Unlike the Earth, the spacecraft won't be surrounded by a blanket of gas to trap heat energy. So there is a danger that the temperature inside the spacecraft will drop too far, killing the astronauts.

Space is about as cold as it is possible to be, with a temperature of about –270°C. Keeping the living quarters at a comfortable 20°C is going to be difficult to achieve.

It won't be possible to stop heat energy flowing completely from the living quarters at 20°C to outer space at –270°C, but there is a lot that can be done to slow down the heat loss. Any heat lost will have to be replaced to keep the inside temperature steady, using up fuel. Extra fuel increases the mass of the spacecraft, making it more difficult and expensive to launch from the Earth.

shiny surface to reduce radiation

photocells make electricity from sunlight

metal foam to reduce conduction

spherical shape reduces surface area for radiation

living quarters at +20°C

stores

space at –270°C

▲ *Possible spaceship design*

Questions

1 Explain why the spacecraft can't lose heat energy by convection or conduction.

2 Why is the best shape of the spacecraft for heat energy management a sphere?

3 Explain why the shiny surface of the spacecraft cuts down heat loss.

4 Explain why the outer hull of the spacecraft is made from a metal foam.

5 Explain, in detail, how the photocells provide heat for the living quarters.

6 Suggest why the stores of fuel and food and water are placed between the living quarters and the outer hull.

7 The photocells will be much colder than the spacecraft. Explain why they are attached to the spacecraft by long thin hollow rods.

Food and phones

In this item you will find out

- about the properties of microwaves and infrared radiation

- how microwaves and infrared transmit energy to materials

- what microwaves are used for

Have you ever worried about what your mobile phone might be doing to your brain? And is the radiation from mobile phone masts a danger to your health?

Mobile phones send out **microwave** signals. These are very low power – much less than those used in microwave ovens. The information carried by the telephone microwaves is transmitted from your phone to a radio mast. To send a signal your phone needs to be able to 'see' the mast. This is because microwaves cannot bend round corners or over buildings. Places that are not in line of sight with a mobile phone mast get poor signals. Mobile phone masts emit a lot more radiation than your phone does.

▲ Some mobile phone masts are disguised as trees

 a Why might people worry about living close to a mobile phone mast?

Microwaves are used every day to cook food. Food can also be cooked in a conventional oven, heated by infrared radiation emitted by hot surfaces. Although it isn't obvious, both infrared and microwaves are similar. Like light, they are both waves carrying energy through the atmosphere. They have the same speed, only different **frequencies**.

This crucial difference decides how much energy they can deliver to an atom or a molecule. Microwaves, with their low frequency, can't deliver enough energy to break a chemical bond. But light, with its high frequency, can break chemical bonds and can therefore be used by plants for photosynthesis. No amount of microwave radiation can make a plant turn water and carbon dioxide into glucose and oxygen – it will just shake the molecules around a bit.

▲ Microwaves from your mobile phone are absorbed by your brain

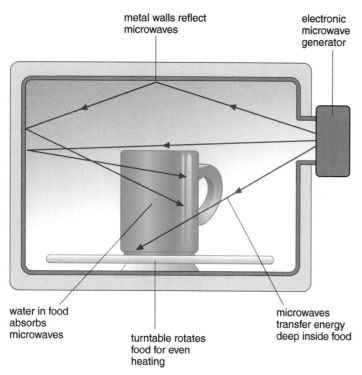

metal walls reflect microwaves

electronic microwave generator

water in food absorbs microwaves

turntable rotates food for even heating

microwaves transfer energy deep inside food

▲ Microwaves heat up the water inside food

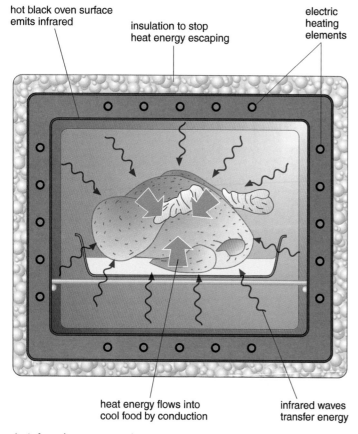

hot black oven surface emits infrared

insulation to stop heat energy escaping

electric heating elements

heat energy flows into cool food by conduction

infrared waves transfer energy

▲ Infrared waves carry heat energy from the hot sides of the oven to the food

Microwave meals

Have you ever wondered how a microwave oven works? When microwave radiation is absorbed by water the water heats up.

Microwave ovens emit microwaves that penetrate about 1 cm into the food. The heat energy is only transferred to water particles in the food, increasing their kinetic energy. The heat energy is then conducted to the cool surface of the food from its hot interior.

Water molecules are special, with opposite charges at each end. This is why they can absorb energy from microwaves. If objects contain no water, then microwaves pass straight through them.

Your body contains a lot of water so you can get burned by microwave radiation if your body tissue absorbs it.

Microwaves cannot go through metal but they can go through glass and plastic. This is why the glass window of an oven has a metal grid to reflect the microwaves back into the oven.

 Why is the inside of a microwave oven made of metal?

Cooking with infrared

Most conventional ovens use infrared radiation to cook food. Lots of infrared is emitted by hot surfaces and more infrared is emitted and absorbed by black surfaces than by any other colour.

Infrared radiation is absorbed by all the particles on the surface of the food, increasing their kinetic energy. This energy is then conducted from the surface to the centre of the food heating it up. Infrared is also reflected by shiny surfaces.

 Why is black the best colour for the inside of an oven?

Energy in waves

Microwaves, infrared and visible light are all electromagnetic waves which carry energy through empty space at the amazing speed of 300,000 km/s. The only difference is their frequency. The frequency of the wave is the number of vibrations in each second. It is measured in hertz (Hz).

high energy red light from a laser

low energy microwaves from an oven

medium energy infrared from a fire

▲ *The different frequencies of microwaves, infrared and visible light*

Cool objects have very little heat energy. They emit low frequency waves. Hot objects have more heat energy, so they emit waves with a higher frequency. If they get hot enough, the frequency becomes so high that the waves are visible as red light.

▲ *Hot objects emit red light*

d **How does the temperature of an object relate to the frequency of its electromagnetic waves?**

The energy delivered to a molecule by an electromagnetic wave depends on its frequency. The higher the frequency, the greater the energy delivered. So high frequency light can break chemical bonds, but low frequency microwaves cannot.

Microwave communications

Diffraction and interference of microwaves used for communications can cause the signal to be weakened. Transmitters need to be high above the ground so that people, buildings and other obstacles do not get in the way of the signal. The transmitters also have to be fairly close together.

◀ *Microwaves for long distance telecommunications are beamed out well above ground level*

keywords

diffraction • frequency • microwave

Could phones give you cancer?

In 1997, a team of scientists in Australia suggested that microwaves from mobile phones could cause cancer. The scientists bred 200 mice genetically disposed to getting cancer of the white blood cells. They then exposed half of them to pulsed digital microwaves close to their heads and waited 18 months for cancers to develop.

They found that mice exposed to the microwaves had 2.4 times as many cancers as the control group. The scientists said that their findings did not mean that human beings ran the same risk as genetically adapted mice. More research was needed.

Of course, you cannot do these sorts of experiments on people. So the scientists will have to compare the cancer rates for people who use mobile phones and for those who do not. They will need to find a large group of people who use mobile phones and then find the same number of people who do not use mobile phones. They will also have to make sure that both groups are matched for sex, age and lifestyle.

This sort of experiment takes many years to deliver any useful data. Until now, no research of this type, even involving many thousands of people, has shown conclusively that using a mobile phone can give you cancer.

Questions

1 Describe how the scientists can find out if mobiles cause brain cancer in humans.

2 Explain why the experiment is best run with a large number of people over a long time.

3 If the experiment eventually suggests a link between mobile phones and cancer, will you stop using one? Give reasons for your answer.

4 One scientist proposes to repeat the mice experiment on cats. Another scientist says the experiment would be better carried out on monkeys. Which scientist do you think is right and why?

5 The frequency of microwaves is a hundred thousand times too low to deliver enough energy to break chemical bonds. Suggest what else might be linking mobile phones to brain cancer?

Everyday infrared

In this item you will find out

- the differences between analogue and digital signals
- the advantages of using digital signals
- what happens inside optical fibres

Sound is an example of an **analogue** signal. The electrical signal from a microphone has a continuously variable voltage.

Infrared can't travel through solid walls, so it is ideal for remote controls. Why can you use two or more different remote controls in the same room? Remote controls use a code to carry information. Each device has its own code, so it ignores infrared signals intended for other devices. The infrared communication sent by a TV remote works best with pulses – the beam of infrared is either on or off. This is an example of a **digital** signal.

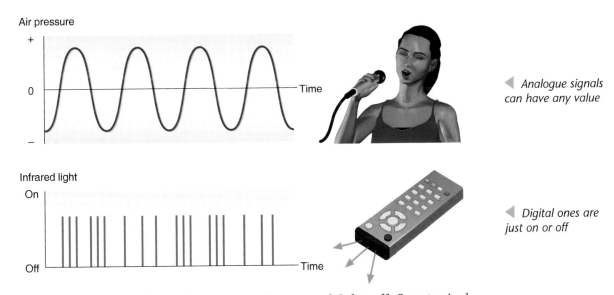

◀ Analogue signals can have any value

◀ Digital ones are just on or off

Digital signals are usually written using 1 for on and 0 for off. So a typical digital signal might look like this on the printed page.

011000110010

If you want to convert analogue signals into digital signals before transmitting them, then you have to use a code.

a A motorcyclist can indicate a left turn with a raised arm or a flashing light. Which is the analogue signal?

Why digital?

Telephone systems send sound messages over long distances. The analogue signal from the microphone is sampled 6,400 times a second, coding the voltage at each instant into a string of eight ones or zeros. For example, 182 mV might be coded as 10110110. A one means the signal is on and a zero means the signal is off.

There are some advantages to having information coded digitally like this. The information is much less likely to be distorted in transmission by noise or interference. Also, more than one signal can be sent at the same time. This is called **multiplexing**.

Amazing fact

A minute of two-way telephone conversation requires about 4 million pulses of infrared.

◀ *Digital coding allows mobile phones to transfer information between people without error*

Optical fibres

Optical fibres are thin glass strands that can stretch for many kilometres. Infrared or visible light signals can pass through them. Pulses of radiation fed in at one end of the optical fibre pass all the way to the other end.

This allows optical fibres to transmit data quickly over large distances. They also allow more than one signal to be sent at a time (multiplexing) and the signals do not interfere with each other.

Even though the optical fibre is very transparent, the radiation can't leak out of the walls. Because the pulses are travelling almost parallel to the wall, they are totally reflected back into the glass. So infrared or visible light has to follow the path of the optical fibre – even round corners! This is called **total internal reflection** (TIR).

b Why doesn't optical fibre need to be coated with a reflector?

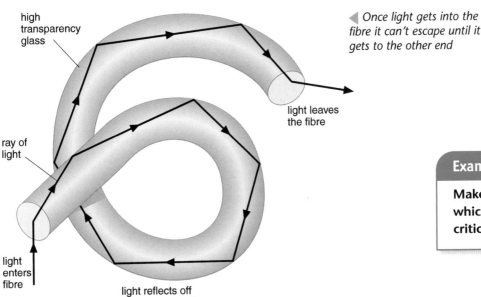

high transparency glass

ray of light

light enters fibre

light reflects off transparent wall

light leaves the fibre

◀ Once light gets into the fibre it can't escape until it gets to the other end

Critical angle

Light is refracted as it passes out of glass into air. Because the light is speeding up as it leaves the glass, the angle of **refraction** is always greater than the angle of incidence.

The angle of refraction cannot be greater than 90°, otherwise the light cannot leave the glass. A beam of light can only pass out of a transparent material (such as water, glass or perspex) if its angle of incidence is less than the **critical angle**. Each material has its own value. For glass, the critical angle is about 40°.

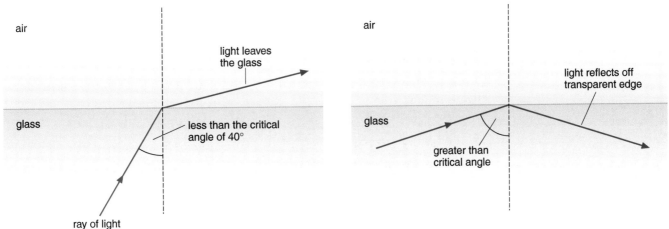

air

glass

ray of light

light leaves the glass

less than the critical angle of 40°

air

glass

light reflects off transparent edge

greater than critical angle

▲ Light can only escape if the angle of incidence is small enough

If the angle of incidence is equal to the critical angle, the light emerges from the material travelling parallel to its surface. When the angle of incidence is greater than the critical angle then the light is internally reflected off the transparent edge to produce total internal reflection.

c The critical angle for seawater is 50°. Describe what happens to a beam of light in the water if it hits the surface at an angle of 60°. What happens if the angle is only 40°?

keywords

analogue • critical angle • digital • multiplexing • optical fibre • refraction • total internal reflection

What's the limit?

A single thin strand of ultra-pure glass, finer than a human hair, can carry about half a million telephone conversations over 100 km without any loss of information. How is this possible?

The transparency of glass varies with the frequency of the light passing through it. At two frequencies, 229 THz and 194 THz, the glass is so transparent that at least 1% of the energy in a pulse is still there after passing through 100 km of fibre. These frequencies are in the infrared, just outside the range of human eyesight.

Although so little of the pulse is still there when it arrives at the other end of the fibre, it can easily be regenerated by an amplifier. Providing that the pulses are regenerated every 100 km, there is no loss of information as it is coded in the timing of the pulses and not their energy.

The amount of data that a fibre can carry is measured in bits per second, where a bit can be 1 or 0. A single fibre can carry at least 10,000,000,000 bits per second and they are grouped in pairs for two-way communication. A single TV channel needs about 20,000,000 bits per second, so one fibre can carry enough information for 500 different channels at any one time.

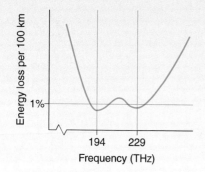

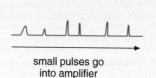

small pulses go into amplifier

amplifier increases energy of pulses without altering their pattern in time

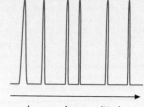

large pulses emitted by amplifier

Questions

1 Explain why long-distance optical fibres use pulses of infrared rather than visible frequencies.

2 Explain why pulses retain their information as they pass down a fibre, even though they lose their energy.

3 A two-way telephone conversation needs a data rate of about 200,000 bits per second. How many conversations could be carried by a pair of fibres, if each fibre can carry 10,000,000,000 bits per second?

Cut the cable

In this item you will find out

- how wireless technology can be used

- how the radiation used for communication can be refracted and diffracted

- how long-distance communications rely on reflection of waves

Cables of wire for ordinary computers are expensive to install and the need to be plugged in keeps you in one place all the time. But wireless laptop computers are becoming more available, with new wireless networks appearing all the time.

This means you can take your laptop anywhere and access the Internet or your emails 24 hours a day. Mobile phones and radios also use **wireless technology**.

All this is possible because wireless technology uses electromagnetic radiation to send and receive data.

But are wireless networks secure? The waves spread out in all directions, so anyone could listen in. However, the power of these transmissions is usually kept low, so they cannot travel far before becoming too weak to pick up.

If signals from different networks do stray into the same area, then careful selection of the frequencies allocated to each network usually avoids confusion for the receiver.

Satellites now allow fast broadband connections to be set up anywhere. A local base station uses a beam of microwaves to exchange signals with a satellite in orbit over the Equator. The satellite has an orbit time of exactly 24 hours, so it appears to be in the same part of the sky all the time, making it easy to beam microwaves to and from it with a fixed dish aerial.

Although placing a satellite in this orbit is expensive, it is often far cheaper than installing optical fibre links across land or sea to remote locations.

a Describe and explain the special orbit needed for communication satellites.

BEFORE WIRELESS NETWORKS

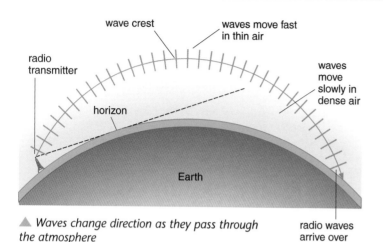

wave crest

waves move fast in thin air

radio transmitter

waves move slowly in dense air

horizon

Earth

▲ *Waves change direction as they pass through the atmosphere*

radio waves arrive over the horizon

Refraction

The radiation used for communication can be refracted (change direction). Waves refract when they change speed. The speed of a wave depends on what it is passing through. So as radio waves, for example, pass through the Earth's atmosphere they are refracted as they move from high density air into low density air, speeding up. This allows radio waves to communicate over the horizon.

b Use ideas of refraction to explain how radio waves from a transmitter mast can be received over the horizon.

Amazing fact

Marconi used the ionosphere to send the first radio signals across the Atlantic in 1901, even though he didn't know it existed.

Ionospheric reflection

Radio waves used for long-distance communication are reflected off a layer of charged particles high up in the atmosphere called the **ionosphere**. This allows distant TV signals to travel long distances, well over the horizon, sometimes interfering with local TV signals. But the height of the ionosphere varies during the day as the atmosphere heat up and cools down, so signal strength can vary considerably!

c With the help of diagrams, explain why over-the-horizon reception of radio waves depends on the time of day.

Satellite communications

High above the Equator is a ring of satellites in geostationary orbits, which means that they stay above the same point on the Earth as it rotates about its axis. Microwaves are used to communicate with these satellites because they travel in straight lines, making it easy to beam them at the satellites.

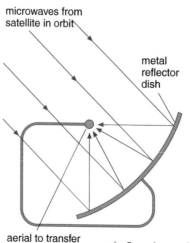

microwaves from satellite in orbit

metal reflector dish

aerial to transfer microwaves to electricity

▲ *Focusing a signal using a reflector dish*

Satellite TV aerials use curved dishes to increase the strength of the signal from the satellite in space. The microwaves reflect off the dish and are focused onto the aerial.

d Transmitters which send radio signals to satellites in space have large round dishes behind the aerial. Why?

Losing the signal

The radiation for long-distance communication can be affected by both refraction and diffraction.

Refraction of waves can occur where different layers of the Earth's atmosphere meet. This can lead to a loss of signal at the receiver. Microwaves are not affected by refraction, which is why they are beamed at satellites rather than radio waves.

Transmitter dishes collect signals from the transmitter aerial to make the beam. The diameter of the dish affects how much the beam spreads out (diffracts) as it travels up to a satellite. Provided that the diameter of the dish is much greater than the wavelength of the waves, the beam won't spread out and the signal will not be lost.

 Why are microwaves used to communicate with satellites instead of radio waves?

Radio reception

Radio stations broadcast their signals from transmitter masts to our homes. Locating the masts in high places increases the range of the radio waves. You can only get good reception if the mast is above your horizon. If you are too far away from the mast, then the signal at your radio will be too weak for you to hear the sound clearly.

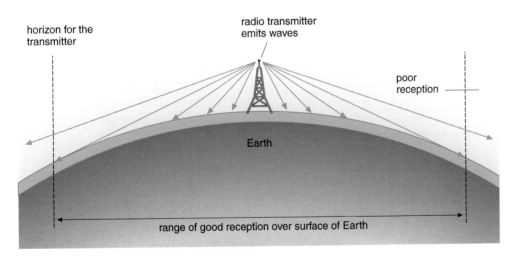

The transmitter emits lots of different radio waves at once, each with a different frequency. Each radio station is allocated its own set of **transmission frequencies** (called a channel) for coding its sound. When you change the station on your receiver, you are changing the range of transmission frequencies that your set is receiving.

 What would happen if two different stations tried to use the same set of transmission frequencies?

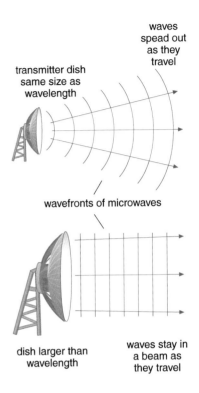

waves spread out as they travel

transmitter dish same size as wavelength

wavefronts of microwaves

dish larger than wavelength

waves stay in a beam as they travel

Examiner's tip

Don't confuse refraction with diffraction.

◀ *A high location for the transmitter increases its range*

keywords

ionosphere • transmission frequencies • wireless technology

Digital radio

Until recently, the highest quality radio broadcasts were in the frequency modulation (FM) band between 88 and 104MHz. Each station in this band broadcasts stereo signals over the whole audible range, using analogue coding.

However, reception is not always perfect and only about a dozen different stations can broadcast in each area around a transmitter.

Because some of the radio signal goes over the horizon, each station is allocated a different channel in adjacent areas. So if you are in a car, you have to change channels to keep a clear signal as you travel.

The sound often flutters as you drive through places where reflected and direct waves overlap, for example near large buildings or near bridges.

Digital Audio Broadcasting (DAB) uses a different method of encoding sound information to get around these problems. It uses digital coding – switching radio signals on and off in the band between 218 and 230MHz.

This improves the reception, with less noise or interference. The sound information is also spread over 1,536 different channels, instead of the one for FM. The same channels are used for each station over the whole of the UK, getting round the problem of signals straying over the horizon from one area to the next. So you don't have to retune your car radio as you travel around.

Finally, any reflections won't disrupt all of these channels, so enough information always arrives at the receiver to assemble the original sound. Best of all, almost a hundred different stations can broadcast in the same area!

transmitter emits waves

large building reflects waves

two sets of waves overlap here and cancel

▲ *Waves can be reflected by buildings which can cause interference*

Questions

1 Describe and explain the disadvantages of broadcasting music in the FM band.

2 What are the advantages of DAB compared with FM broadcasting?

3 Explain how DAB performs better than FM.

4 Explain why reflections of radio waves can lead to poor reception.

Making light work

In this item you will find out

- the main features of a wave

- how pulses of light can be used to send messages in code

- how a laser reads digital information from a CD

▲ *You can use a torch to send light signals*

Light has been used to send messages from place to place for centuries. Fire beacons used to be lit to warn people of invaders, most famously in 1588 when the Spanish Armada arrived in English waters.

In 1832 Samuel Morse invented his famous code of dots and dashes.

Morse code was first used to send messages through wires by pulsing the current in them on and off. This telegraph system was an early version of the Internet, allowing quick and secure communication across the whole surface of the Earth for the first time.

Pulses of light can also use Morse, and other types of code, to send messages. Armies and navies quickly adopted this technology, for use where wires were impractical.

Even when radio waves were discovered and used for sending messages in Morse code, pulsed light continued to be used because it could be beamed. You could only receive the message if you were in the path of the beam, which made it more secure.

Amazing fact

Light takes eight minutes to get from the Sun to the Earth.

A •– A	N –• N	0 –––––0
B –••• B	O ––– O	1 •––––1
C –•–• C	P •––• P	2 ••–––2
D –•• D	Q ––•– Q	3 •••––3
E • E	R •–• R	4 ••••– 4
F ••–• F	S ••• S	5 ••••• 5
G ––• G	T – T	6 –•••• 6
H •••• H	U ••– U	7 ––••• 7
I •• I	V •••– V	8 –––•• 8
J •––– J	W •–– W	9 –––––• 9
K –•– K	X –••– X	
L •–••• L	Y –•–– Y	
M –– M	Z ––•• Z	

 Suggest one advantage of using light instead of wires to send messages in Morse code.

Light, radio and electricity

How fast can information travel? At the speed of light! You can now get a message to anywhere on Earth in at most a few seconds. You can use radio waves, electrical pulses in wires or light in optical fibres. Since all electromagnetic waves move at the same speed, it doesn't matter which one you use. They all carry information at the same speed. But the amount of information that a wave can carry is determined by its frequency. High frequency signals, such as light, can transfer much more information per second than low frequency ones, such as radio waves.

 Compare the advantages and disadvantages of communicating by electrical signals, light and radio waves in a battle at sea.

Wave properties

Electromagnetic waves are **transverse waves** which vibrate up and down.

The diagram below shows what a transverse wave looks like. It is made up of a series of **crests** and **troughs**.

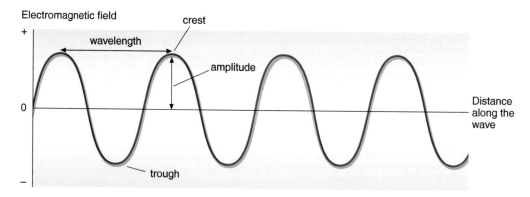

▲ *The features of a transverse wave*

- The frequency is the number of waves produced by the source each second.
- The **wavelength** is the distance from one wave crest to the next.
- The **amplitude** is the height of the crests.

The frequency of a light wave is measured in hertz (Hz). The wavelength is measured in metres. You can calculate the speed of a wave by using this equation:

$$\text{speed} = \text{frequency} \times \text{wavelength}$$
$$c = f\lambda$$

where c = the speed of the wave (m/s)
$\quad\quad f$ = the frequency (Hz)
$\quad\quad \lambda$ = the wavelength (m)

All electromagnetic waves travelling in empty space have the same speed. The speed is always 300,000 000 m/s. This means that if you increase the frequency of a wave, the wavelength will automatically decrease.

For example, what is the frequency of microwaves whose wavelength is 3 cm?

$$c = f\lambda$$

$$300,000,000 \, \text{m/s} = f \times 0.03 \, \text{m}$$

$$f = \frac{300,000,000}{0.03}$$

$$f = 10,000,000,000 = 10 \, \text{GHz}$$

c   FM radio broadcasts have a frequency of about 100 MHz. What is their wavelength?

d   What happens to the frequency of an electromagnetic wave when its wavelength is doubled?

Lasers

Lasers produce a special type of light. The beam is narrow and very intense. The waves all have the same frequency and are all in phase with each other.

Compact disks use a digital code to store information about the music recorded on them. The code, which is stored as a spiral of pits in a disc of plastic, is read by a laser. Each time a pit passes below the laser, the light reflected from a shiny surface onto a detector decreases. Lots of reflection corresponds to a 1, only a little corresponds to a 0. About 1,000,000 bits of information are needed to store one second's worth of music!

e   Describe how a laser reads CDs.

Lasers can also read bar codes on products in supermarkets and link computers with optical fibres.

Measuring Mars

In 1998 a spacecraft was placed in circular orbit around Mars. The robot spacecraft (known as the Mars Global Surveyor) had several instruments on board. One of them, the Mars Orbital Laser Altimeter (MOLA), spent the next two years producing detailed maps of the surface of Mars. These maps were crucial to the success of later robot missions.

MOLA makes its measurements as follows:

- A laser emits a very short pulse of light towards the surface.
- The pulse travels about 400 km to the surface of Mars.
- Some of the pulse is reflected off the surface and sets off back to MOLA.
- MOLA detects the reflected pulse.

MOLA measures the time elapsed between the emission and reception of each laser pulse.

Knowing that light travels at 300,000,000 m in each second, it is a simple matter work out the distance between MOLA and the surface.

In this way, the surface of Mars has been mapped to a precision of 10 m.

Some features of laser light are crucial for MOLA. The light spreads out very little as it travels – each pulse has a width of only 100 m when it hits the surface of Mars.

The light pulse is also very short – a billion pulses can happen in each second. A lot of energy can be put into each pulse and the waves are in phase with each other.

Questions

1 Describe and explain the function of MOLA.

2 MOLA is about 400 km above the surface of Mars. If light travels at 300,000 km/s, how long does it take each pulse to go from MOLA to Mars and back?

3 Describe the properties of a laser which are crucial to the operation of MOLA.

4 Explain why Mars Global Surveyor transmits its signals to Earth in digital form through a large dish aerial.

Killer waves

In this item you will find out

- how earthquakes produce seismic waves

- about the dangers of exposure to ultraviolet radiation

- how humans are affecting weather patterns

The Sun emits **ultraviolet** radiation. The short wavelength of ultraviolet radiation (UV) means that it has enough energy to smash chemical bonds.

It can easily damage molecules (such as DNA) in living cells, causing them to divide rapidly and become independent of the other cells around them. So live skin cells exposed to ultraviolet can develop cancer.

This is why many people in sunny parts of the world have dark skins to absorb the ultraviolet and prevent it from reaching body tissues under the skin.

What happens if you have a pale skin? Given enough time, your skin will turn a darker colour when exposed to ultraviolet radiation. Sunburn is a painful sign that ultraviolet is damaging you.

Fortunately, a layer of **ozone** high up in the atmosphere absorbs a lot of the Sun's ultraviolet radiation.

Ozone is O_3, caused by the action of UV on oxygen O_2. However, there is evidence that man-made chemicals called CFCs, which contain chlorine, are destroying the ozone layer.

a   **Explain how the ozone layer protects us from skin cancer.**

▲ *Dark skins absorb more ultraviolet than pale skins*

Amazing fact

UV is used to manufacture vitamin D in your skin. White skins evolved because of the need to make this vitamin in a northern climate where sunlight is weak.

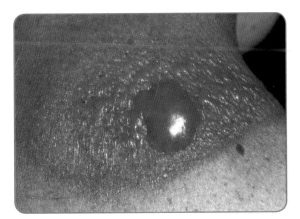

◀ *Too much exposure to UV can cause skin cancer*

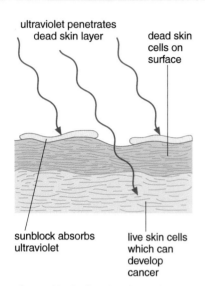

ultraviolet penetrates
dead skin layer

dead skin
cells on
surface

sunblock absorbs
ultraviolet

live skin cells
which can
develop
cancer

▲ *Sunblock absorbs ultraviolet
before it reaches live cells*

Sunblock

You can avoid sunburn by putting sun block on exposed skin. The **sun protection factor** (SPF) of a sunblock indicates how much UV it absorbs. For example, an SPF of 8 allows you to stay up to eight times longer in the sun without burning than if you were not wearing any sunblock at all.

 Julie usually burns in the hot sun after 20 minutes. How long can she stay in the sun if she puts on sunblock with SPF 30?

Climate change

Two things keep our Earth at a comfortable temperature:

- The Earth absorbs light and infrared from the Sun, heating it up.
- The Earth radiates infrared into space, cooling it down.

Humans and nature can both alter this delicate balance, causing the temperature of the planet to change gradually:

- Dust from volcanoes reflects radiation from the Sun which cools down the Earth.
- Dust from factories reflects radiation from cities which causes warming.
- Using more fossil fuels means creating more CO_2, which stops infrared radiation escaping from the Earth and warming it up.
- Burning forests also means creating more CO_2, which causes the Earth to warm.

A warmer planet means that sea levels will rise as the sea warms up and expands. As the temperature changes, the climate will change, which may lead to the extinction of some animals and plants.

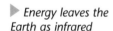 Explain why burning fossil fuels is changing our climate.

▶ *Energy leaves the Earth as infrared*

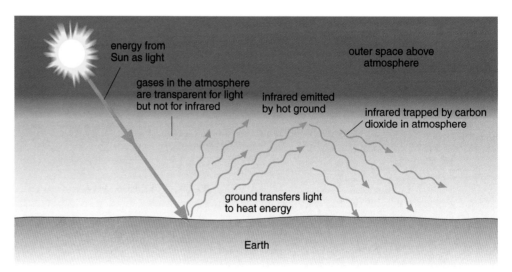

energy from
Sun as light

outer space above
atmosphere

gases in the atmosphere
are transparent for light
but not for infrared

infrared emitted
by hot ground

infrared trapped by carbon
dioxide in atmosphere

ground transfers light
to heat energy

Earth

Violent Earth

Earthquakes suddenly release a lot of energy deep underground. The energy is carried away by seismic shock waves which can travel inside the Earth. One type of wave, called a **P-wave**, pushes the ground back and forth as it moves along. It is a **longitudinal** wave.

The other type of wave shakes the ground from side to side. It is called an **S-wave** and it is a transverse wave. When they reach the surface, both of these waves can cause a lot of damage.

S-waves can travel through solids but not through liquids. P-waves can travel through solids and liquids. Because P-waves can travel across the liquid core of the Earth and through all the layers of solid and liquid rock, they can be detected by seismometers on the other side of the Earth. P-waves also travel faster than S-waves.

Seismometers are located in many places on the Earth's surface. By comparing the times at which the S- and P-waves from an earthquake arrive at each seismometer, scientists have been able to construct a theory of the structure of the Earth. This is because the two types of waves have different properties.

▲ A trace produced by an earthquake

◄ S-waves and P-waves have different effects on the ground that they pass through

moves up and down

surface

P-wave pushes and pulls as it travels

epicentre

moves from side to side

surface

S-wave shakes from side to side as it travels

epicentre

d An earthquake happens deep under your feet. What will the P-waves do to you? What will the S-waves do to you?

---

keywords

earthquake • longitudinal • ozone • P-wave • sun protection factor • S-wave • ultraviolet

Climate modelling

Predicting the future climate of the Earth is a tricky business. Almost every week, scientists publish the findings of their research and reach a different conclusion. Although there is now general agreement that the Earth's surface is getting hotter, there is still much argument over the cause and implications.

In Indonesia, for example, there are fires smouldering uncontrollably in peat seams just under the Earth's surface. A fossil fuel is burning, so this raises the level of carbon dioxide in the atmosphere allowing it to heat up. The fires also put a lot of smoke into the atmosphere. This reflects some sunlight back into space before it is absorbed by the Earth and so cools it down. Which effect dominates – the cooling or the heating?

There are other difficulties faced by scientists trying to predict future climate. Warmer oceans result in more evaporation of water from the sea. This creates more clouds. But clouds are white and reflect sunlight, cooling the Earth down. Once again, a warming also results in a cooling.

It is possible that these examples will keep our climate stable whatever we do to the atmosphere. However, there are plenty of situations where warming means more warming, for example the melting of glaciers in the Alps. As the ice melts, white reflecting snow is replaced with black absorbing rock. A warmer Earth results in more warming, so the ice melts more rapidly, the warming becomes more rapid – until all of the ice has melted.

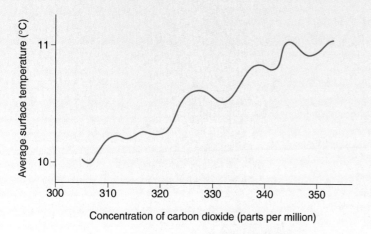

Questions

1 Explain what effect burning peat has on the climate.

2 Explain why melting ice caps might result in a warmer Earth.

3 Does the fact that surface temperature increases with increasing carbon dioxide levels provide scientific proof that man's activities are affecting the climate? Suggest what other evidence you would need.

4 There is a lot of frozen peat underground in Northern Europe and Asia. When peat unfreezes, it decomposes into methane, a potent greenhouse gas. Explain what will happen to global warming when that peat gets warm enough.

P1a

1 Complete the sentence.

The specific heat capacity of a material is the ____(1) needed to raise the temperature of one ____(2) by one ____(3)   [3]

2 Explain what the specific latent heat of material is. [2]

3 A hot object is placed in contact with a cold one.

a Which way will energy flow between them? [1]

b What will that flow of energy do to the temperature of the objects? [3]

4 Heat and temperature are two quantities that can be measured for an object.

What do these quantities tell you about the object? [2]

5 Calculate the energy required to raise the temperature of 5 kg of water by 25 °C. The shc of water is 4,200 J/kg/°C. [3]

6 3.0 kJ of energy are required to melt 15 g of a sample of fat.

a Explain why the energy does not raise the temperature of the fat. [1]

b Calculate a value for the specific latent heat of the fat. [3]

7 Complete the sentences.

Chocolate melts at 32 °C. Your mouth has a ____(1) of 37 °C. When solid chocolate is put in your mouth, its temperature ____(2) as it absorbs ____(3) energy from you.

When the chocolate reaches its melting point of ____(4), its ____(5) stops rising as it absorbs ____(6) heat and turns from a solid to a ____(7).

The liquid finally reaches a temperature of ____(8) as you swallow it. [8]

P1b

1 Insulating the roof of a house costs £1,000, saving £200 a year in heating bills.

Calculate the payback time for insulating the roof. [2]

2 Some meat is being cooked in a microwave oven. For every 60 J of electricity transferred into the oven, only 45 J of heat energy transfers into the meat. Use the rule:

$$\text{efficiency} = \frac{\text{useful energy output}}{\text{total energy input}}$$

to calculate the efficiency of the oven. [2]

3 Use ideas of heat transfer to explain why a cooking pan has a metal base but a wooden handle. [2]

4 Heat energy is transferred out of houses by conduction, convection and radiation.

Describe one example of each of these processes. [3]

5 Describe two ways of reducing heat energy loss by conduction in a house. [2]

6 Describe two ways of reducing heat energy loss by radiation in a house. [2]

7 A fridge claims to have an efficiency of 0.85. For every 200 J of electricity put into the fridge, how much of that energy does something useful? [2]

8 Complete the sentences.

Heat energy is lost through the solid walls of a house by ____(1).

As warmed air rises from the roof, heat energy is lost through ____(2).

Increasing the temperature inside the house increases the amount of heat lost by ____(3) through the windows. [3]

P1c

1 Match the start and end of the sentences.

1 a cavity wall	a reduces heat loss by conduction
2 drawing curtains across windows	b reduces heat loss by radiation
3 planting trees around a house	c reduces heat loss by convection [2]

2 Insulating materials are usually full of holes. Explain why these holes make the materials poor conductors of heat. [2]

3 Heat energy can be saved by painting a house white. State and explain the type of heat transfer that this reduces. [2]

4 Describe and explain how convection currents transfer energy across a cavity wall in a house. [3]

5 Heat energy is transferred through solids by conduction.

a Use ideas about particles to explain conduction. [2]

b Explain how the structure of solids can be modified to reduce conduction. [2]

6 Explain why reducing the interior temperature of a house can reduce the rate at which it loses energy. [2]

7 A pot of ice cream is wrapped in plastic bubble wrap before being surrounded by aluminium foil.

Explain how this helps to keep the ice cream cold in a warm room. Your answer should include: the direction of heat flow; the effect of the foil; the effect of the bubble wrap. [3]

P1d

1 Complete the sentences.

Microwaves are reflected by ____(1) but transmitted by ____(2).

In an oven, microwaves are absorbed by____(3) in the food. [3]

2 Microwaves carry telephone messages from one place to another.

Describe and explain the limits on how far apart the two places can be. [3]

3 Compare the different ways in which food is heated in a microwave oven and a gas oven. [3]

4 The microwave signals which carry long distance telephone messages suffer signal loss due to diffraction.

Explain what this means. [2]

5 Use ideas of particles to explain how food is heated in a microwave oven. [3]

6 It has been suggested that microwaves from mobile phones do not give people cancer of the brain.

Describe how scientists could collect evidence to support this suggestion.

7 A gas oven is a metal box suspended in another metal box.

a Explain why the space between the boxes is packed with fluffy wool made from rock. [2]

b Explain why the surface of the inside box is coloured black. [2]

c Explain why the surface of the outside box is coloured white. [2]

P1e

1 Optical fibres allow the rapid transmission of data using pulses of light.

a Describe how light passes from one end of a fibre to the other. [2]

b What type of signal are the pulses of light? [1]

c The data being transmitted is speech. What type of signal is speech? [1]

2 Light in glass hits the surface below the critical angle. Which of the sentences best describes what happens to the light.

A All of it passes out of the glass into the air.
B Some passes into the air, the rest is reflected back into the glass.
C All of it is reflected back into the glass. [1]

3 Long-distance communications use pulses of infrared along optical fibres.

a Explain why the use of pulses allows almost error-free transmission of data. [2]

b Explain how multiplexing increases the data transmission rate. [2]

4 Light is only totally internally reflected if it approaches the surface of a material at more than the critical angle.

Explain the meaning of this sentence by drawing some ray diagrams. [3]

5 Complete the sentences.

Optical fibres are made of ____(1) which is ____(2) to infrared.

This allows the infrared to travel a long ____(3) before it is completely ____(4).

The infrared hits the edge of the fibre at more than the ____(5) angle, so it is all ____(6) and none can ____(6) out of the glass. [7]

6 Data is sent down optical fibres in digital form.

a Give another example of a digital signal. [1]

b Give an example of an analogue signal. [1]

c What are the advantages of sending data in digital form rather than analogue form? [2]

P1f

1 Name three common uses of wireless technology. [3]

2 Radio waves travelling away from the Earth can sometimes follow a curved path back to Earth.

What is the name of this effect? [1]

3 Why are different radio stations in an area not allowed to broadcast with the same frequency? [1]

4 Satellites in orbit around the Earth are used to relay radio signals so that they can be used for long-distance communication.

State and explain two other ways in which radio signals can be received out of sight of the transmitter. [4]

5 Microwaves for communicating with satellites are formed into a beam by a large reflecting dish.

Why is the size of the dish important? [3]

6 What is the difference between refraction and reflection?

Give an example of each as part of your answer.

P1g

1 A sound wave with a frequency of 680 Hz has a wavelength of 0.5 m. Use the rule

speed = wavelength × frequency

to calculate the speed of the wave. [2]

2 Match the start and end of the sentences.

1	the frequency of a wave is	a	the height of a crest
2	the amplitude of a wave is	b	the distance between troughs
3	the wavelength of a wave is	c	the number of waves in a second [2]

3 Describe how light can be used to send messages in Morse code. [2]

4 Lamps, LEDs and lasers all emit light. What is different about the light from a laser? [2]

5 Explain how a laser is used to extract information from the surface of a compact disc (CD). [3]

6 The wavelength of red light from a laser is 6.3×10^{-7} m.

Calculate its frequency. The speed of light is 3.0×10^8 m/s. [3]

7 Ships can use strings of flags to send messages to each other. They can also use beams of light.

a Explain how a message can be sent along a beam of light. [2]

b Suggest why strings of flags are not used when the ships are part of a naval battle. [1]

c Ships in battle use radio waves for communication instead of light. Suggest why. [1]

P1h

1 Complete the sentences.

There are two types of seismic wave, called ____1) and ____(2). They are produced by earthquakes. ____(3) are transverse and can only travel through ____(4).

____(5) are longitudinal, so they can travel through both solids and ____(6). [6]

2 Explain why people with darker skins have less risk of skin cancer. [3]

3 A sun cream has an SPF of 4. What does this mean? [2]

4 Explain how the action of volcanoes can result in global cooling. [2]

5 Explain how the passage of seismic waves through the Earth provides evidence of its structure. [4]

6 Why are people concerned about environmental pollution from CFCs? [3]

7 Complete the sentences.

Light energy from the ____(1) passes straight through our atmosphere.

It is absorbed by the ____(2) and transferred to ____(3) energy.

The warmed ground emits ____(4) which is partly reflected by ____(5) in the atmosphere.

Increasing the amount of carbon dioxide results in ____(6) infrared from the ground being trapped, raising the ____(7) of the atmosphere. [7]

8 The sentences describe some changes caused by an increase in the temperature of the Earth's surface. Put them in order.

A The surface of the Earth gets warmer.
B The surface of the Earth cools down again.
C The water condenses in the atmosphere to form clouds.
D The rate of evaporation of water from the sea increases.
E Sunlight reflects back into space from the top of the clouds. [4]

P2 Living for the future

I've heard that global warming is going to wreck the Earth and cause massive extinctions. How can I get electricity without being part of this disaster?

Nuclear power doesn't cause global warming. Why aren't we building more nuclear power stations to provide electricity when oil and gas run out?

How can we be so sure that the Sun will still be shining in a billion years time?

- At present, most of our electricity is made by burning fossil fuels. This can't go on for much longer, as these fuels will run out soon. Perhaps this is a good thing, as the gases produced by burning them are causing the surface of our planet to get warmer, with dire consequences for our environment.

- Electricity can be made directly from sunlight, or indirectly with wind turbines and hydroelectric schemes. It can also be made by splitting uranium atoms but this leaves behind a lot of dangerous radioactive waste.

- One thing is certain. The problem of electricity supply won't be solved by emigrating to another planet. There is nowhere else to go. Fortunately, the Sun will carry on flooding us with large amounts of energy for a few billion years more.

What you need to know

- Non-renewable fuels will not be replaced once they have all been used up.

- Electricity in a circuit transfers energy through metal wires.

- Magnets have north and south poles which attract magnetic materials.

- Magnets are surrounded by fields which can interact with other magnets.

Sunlight and wind power

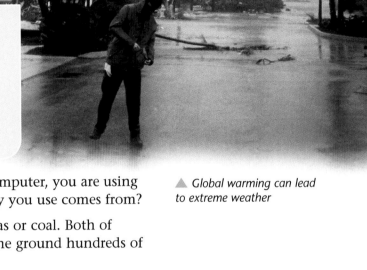

In this item you will find out

- how photocells make electricity from sunlight

- how solar heaters make heat energy from sunlight

- how wind turbines make electricity from wind

Every time you switch on a light, or turn on your computer, you are using energy. Do you ever think about where all the energy you use comes from?

▲ *Global warming can lead to extreme weather*

Most of the electricity you use is made by burning gas or coal. Both of these are fossil fuels, made from carbon trapped in the ground hundreds of millions of years ago.

Burning them restores that carbon to the atmosphere as carbon dioxide, trapping infrared radiation emitted by the ground. So each time you plug your mobile phone into the mains to recharge it, you are heating up the Earth.

Many people are convinced that this increase of carbon dioxide is bad for the planet as it may lead to global warming.

We prefer to buy our electricity from the cheapest sources. At the moment, this means power stations which use gas, coal and nuclear fuel.

Many governments around the world have agreed that this must stop, as it is damaging the environment. The race is on to find alternative ways of making electricity which are also cheap enough to challenge the supremacy of coal and gas.

▲ *This tanker is carrying gas to the UK*

a Explain why so much electricity is made from fuels which pollute the environment.

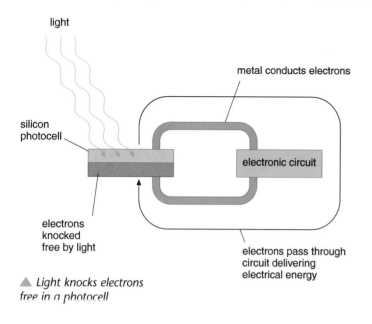

light

metal conducts electrons

silicon photocell

electronic circuit

electrons knocked free by light

electrons pass through circuit delivering electrical energy

Light knocks electrons free in a photocell

Photocells

Photocells convert light energy into electricity. They are made from silicon. The light absorbed by the photocell has enough energy to separate electrons from atoms in the silicon crystals. The electrons can then move freely and as they pass through electric circuits, they transfer their energy to the components.

Photocells make DC (**direct current**) electricity. This is a current that flows in the same direction all the time.

The electrical energy per second (power) from a photocell depends on two things:

- the intensity of the light
- the size of the surface area that can absorb the light.

b Explain how a photocell makes electricity from light.

c Describe and explain how the power from a photocell in the open air changes over 24 hours.

Photocells can be used to power satellites

Pros and cons of photocells

There are lots of advantages to using photocells to make electricity:

- They are easy to maintain.
- They do not need power cables so they can be used in remote locations.
- They do not need fuel as they use a renewable energy resource.
- They are strong and they last a long time.
- They do not pollute the environment.

The one disadvantage is that photocells cannot produce electricity when there is no light – either at night or when the weather is bad.

d Explain why photocells are useful for making electricity in remote locations.

Solar cooking

Many people in the world burn wood to cook their food. Using the Sun instead makes good sense. But you need a curved mirror to focus the light onto the food. You also need to adjust the mirror continuously, so that as the Sun moves across the sky the light stays focused on the food.

e Why are solar powered barbecues not popular in the UK?

f Suggest why people who use solar cookers are advised to wear sunglasses.

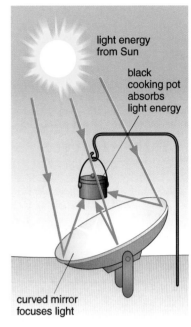

light energy from Sun

black cooking pot absorbs light energy

curved mirror focuses light

▲ *The mirror focuses the light onto the cooking pot*

Solar heating

The Sun's energy can also be used to provide passive solar heating for buildings. You can have a wall of glass on the sunny side of your house with a space between it and the house wall. Solar collectors need to absorb as much sunlight as possible so they are usually positioned on the south side of buildings.

g Explain how the sheet of glass helps to transfer light energy from the Sun into heat energy for the house.

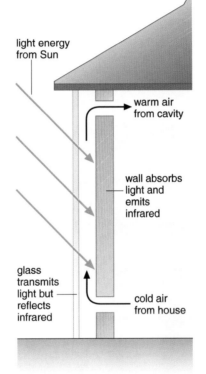

light energy from Sun

warm air from cavity

wall absorbs light and emits infrared

glass transmits light but reflects infrared

cold air from house

▶ *Infrared is trapped between the glass and the wall, heating up the air*

Wind power

A **wind turbine** transfers the **kinetic energy** of moving air in the atmosphere to electrical energy. The amount of electricity a wind turbine produces depends on the wind speed. Wind power is a renewable energy resource, as we will never run out of wind and it does not produce any polluting waste. Wind turbines are rugged but they can take up a lot of space and spoil the scenery.

h Suggest why the most cost-effective wind turbines are large and sited on high ground.

i State two advantages and two disadvantages of using wind turbines to generate electricity.

keywords

direct current • kinetic energy • photocells • wind turbine

Energy for free?

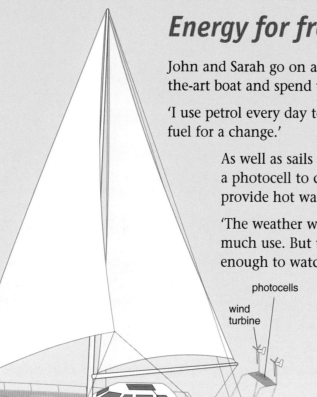

John and Sarah go on a holiday that changes their lives. They hire a state-of-the-art boat and spend two weeks sailing on the sea.

'I use petrol every day to travel to work,' says John, 'I wanted to travel without fuel for a change.'

As well as sails to catch the wind, the boat has a wind turbine and a photocell to charge up the battery. There is even a solar panel to provide hot water.

'The weather wasn't very good,' says Sarah, 'so the photocell wasn't much use. But the strong wind kept our battery topped up nicely – enough to watch TV every evening and keep the lights and navigation equipment going.'

photocells

wind turbine

solar heating panels

John is surprised at how much hot water is available on the boat. There was plenty for hot showers as well as washing dishes.

When they get home, John and Sarah decide to adapt their house to take advantage of free energy from the Sun.

Energy from the Sun and wind is free. But you have to pay for the equipment needed to transfer the energy. So John and Sarah need to decide what they can afford.

They install solar heating panels on the roof and build a glass conservatory on the sunny side of the house. These are quite cheap and pay for themselves after about ten years.

▲ Your house can be heated using energy from the Sun

Questions

1 Explain the four different ways in which the boat uses energy from the environment.

2 Suggest why the boat must be equipped with a diesel generator.

3 Suggest and explain four changes to their house which will allow them to get energy from the Sun and wind.

4 Explain why photocells on their roof would be almost useless without a battery.

5 John and Sarah decide not to install a wind turbine on their roof. Instead, they decide to buy their electricity from a company that uses turbines out at sea. Suggest five reasons why this is a good idea.

Making electricity

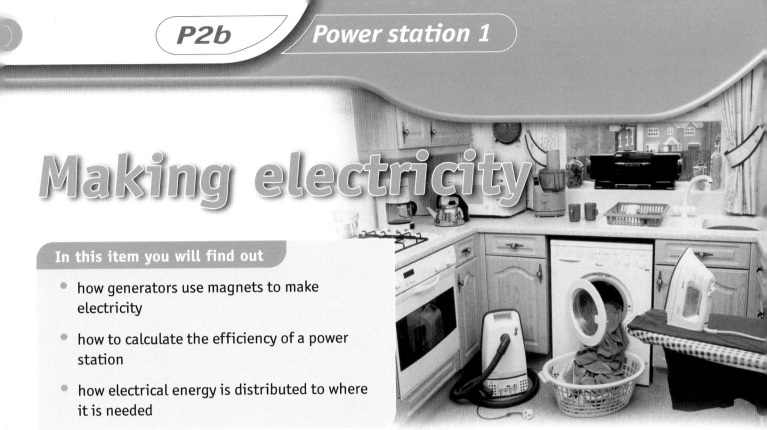

In this item you will find out

- how generators use magnets to make electricity

- how to calculate the efficiency of a power station

- how electrical energy is distributed to where it is needed

Most of us are surrounded by electrical appliances which run off electricity. Here are some of the things they do for us:

- preserve and cook our food
- heat our houses
- clean our floors
- entertain us with sounds and pictures
- carry our messages to other people.

Electricity is everywhere. At the flick of a switch, a host of different useful devices come to life. But where does it come from? What environmental price are we paying for it?

Most of our electricity is generated by power stations that burn fossil fuels. The electricity is then transported around the country and to houses, shops and factories by the **national grid**.

Electricity is carried away from power stations along cables called transmission lines. The transmission lines carry electricity at a very high voltage. An increased voltage means a decreased current, so the transmission lines are heated up less, which reduces energy waste and keeps down costs.

Before a power station can feed its electricity into the grid, it has to be passed through a **transformer** to increase its voltage to several hundred thousand volts. Another transformer between the grid and the consumer reduces the voltage to a safer 230 V.

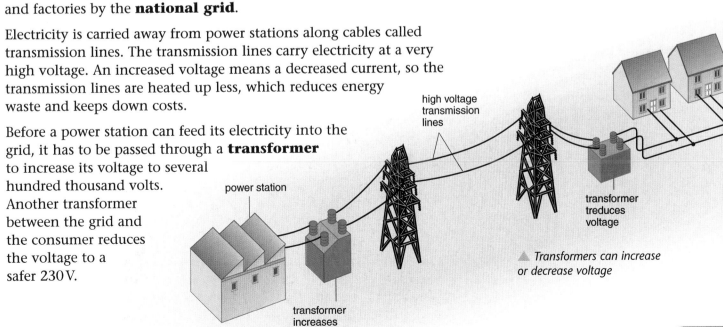

power station

high voltage transmission lines

transformer treduces voltage

transformer increases voltage

▲ *Transformers can increase or decrease voltage*

Magnetism makes electricity

You can generate electricity by either moving a magnet past a coil of wire or moving a coil of wire past a magnet. This is called the **dynamo effect** and produces a voltage in the coil of wire.

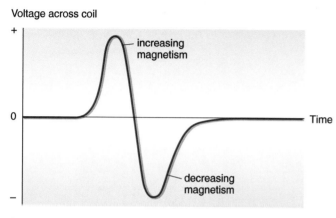

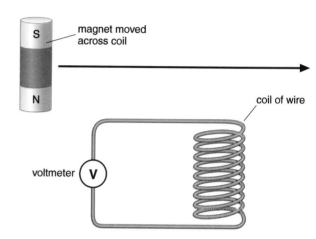

▲ When the magnet is moved past the coil a voltage is produced

The size of the voltage can be increased by:

• having more coils of wire
• wrapping the coil around a magnetic material, such as iron
• moving the magnet or the coil faster
• using a stronger magnet.

Generators

A simple **generator** uses a spinning magnet to transfer kinetic energy into electrical energy. It is made from a coil of wire wound round an iron core. As the magnet spins, it alters the magnetic field in the iron core. Each time the magnetic field changes, a voltage appears across the coil of wire. Iron is used for the core because magnetic fields travel easily through it.

 Describe four ways of increasing the voltage across the generator coil.

Large generators use electromagnets. These are made with coils of wire wrapped around cylinders of iron. The electricity for the electromagnets comes from small generators with spinning permanent magnets made from steel.

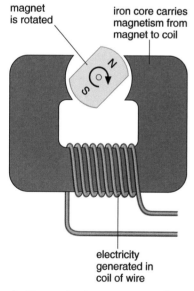

▲ How a simple generator works

Alternating current

The voltage from a generator is always changing. It rises as the magnetism increases, but then has to fall again as the magnetism decreases. So generators produce **alternating current** (AC). The voltage from a battery is different. It has a steady value, producing direct current (DC).

Alternating current from a generator has a frequency. This is the number of cycles of electricity produced in a second. Each cycle is produced by a single rotation of the magnet, so the frequency is fixed by how fast the magnet is rotating. Doubling the speed of rotation not only doubles the frequency of the AC, it also doubles its voltage.

Voltage

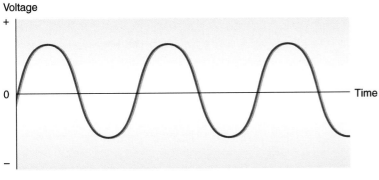

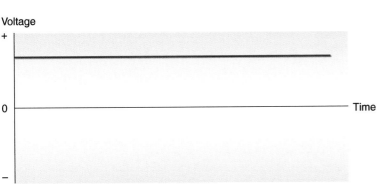

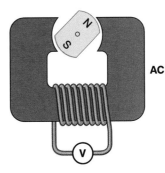

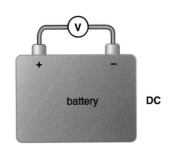

AC

battery   DC

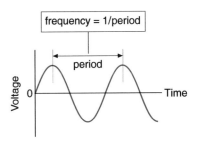

◀ *Voltage–time graphs for a generator and a battery*

b The magnet of a generator spins at 3,000 revolutions per minute, producing one cycle of electricity per revolution. Calculate the frequency of the AC it produces.

Inside a power station

Power stations extract energy from their fuel as heat energy.

- The heat energy is transferred to water, making it boil.
- Heat energy is transferred to strain energy in high pressure steam.
- Steam passes through a turbine, forcing the shaft to spin round.
- The generator transfers the kinetic energy of the shaft into electrical energy.

c How does energy in fuel become electricity?

Not all of the energy in the fuel becomes electricity. Some of it is wasted as heat energy. The efficiency of a power station tells you how well it uses the energy in the fuel. For example, a typical coal-fired power station produces 60 J of waste heat energy for every 40 J of electricity it makes. So how efficient is this power station?

You can find out the energy input from the fuel using the following equation:

fuel energy input = waste energy output + electrical energy output

fuel energy input = 60 J + 40 J = 100 J

You can then calculate the efficiency of the power station:

efficiency = electrical energy output/fuel energy input

efficiency = 40 J/100 J = 0.4

d A gas-fired power station wastes 80 J of heat energy for every 120 J of electricity it makes. What is its efficiency?

frequency = 1/period

period

Voltage

0   Time

▲ *A voltage–time graph for AC*

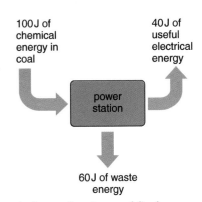

100 J of chemical energy in coal

40 J of useful electrical energy

power station

60 J of waste energy

▲ *Energy flow in a coal-fired power station*

keywords

alternating current
• dynamo effect •
generator • national grid
• transformer

▲ *Nicolai Tesla in his laboratory*

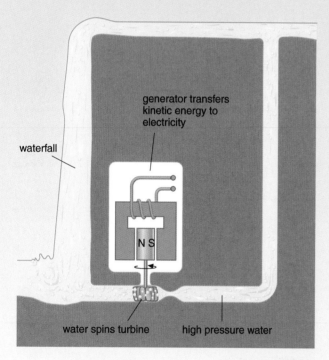

generator transfers kinetic energy to electricity

waterfall

N S

water spins turbine

high pressure water

Tesla's triumph

It is 1893. George Westinghouse is awarded the first contract to make electricity from the Niagara Falls, where huge quantities of water tumble down a sheer 50 m drop. The kinetic energy of the falling water is wasted as heat and sound energy. Westinghouse plans to transfer some of that energy into electricity.

The power station costs a lot of money to build. To make that investment pay off, Westinghouse needs to sell lots of electricity to a large city, such as New York.

Unfortunately, New York is about 300 km away from Niagara Falls. Most of the electricity fed into the transmission lines will be transferred to heat before it arrives at the other end. In 1893, New York used DC electricity, generated by many small coal-fired power stations in the city.

Westinghouse invests in new, unproven technology. He appoints Nicolai Tesla, the inventor of the AC generator and AC motor, as his chief engineer. Tesla already knows that the key to an efficient transmission line system is low current in the cables. The equation $P = VI$ tells him that, for the same power to be transmitted, small currents can be achieved if the voltage is high. He also knows that transformers can be used to change the voltage of AC, with an efficiency close to 1.0.

So he installs AC generators at the Niagara Falls and by 1896 is able to supply New York with cheap electricity. Within ten years the old local DC generators are dismantled and the whole world adopts Tesla's methods.

Questions

1 Describe the energy transfers for Tesla's Niagara Falls power station.

2 Why did New York need lots of DC power stations?

3 Explain why raising the voltage of a transmission line reduces the current in it. Why is this a good thing to do?

4 Tesla placed a transformer at each end of the transmission line. What did each transformer do?

5 A particular transmission line system has an efficiency of 0.92. If the power transferred is 500 MW, how much power is wasted as heat in the cables?

Fuel for electricity

In this item you will find out

- about the advantages and disadvantages of different energy sources

- about the cost of using electricity

- some benefits and drawbacks of nuclear power stations

There are many ways of heating water to make steam in a power station. We can:

- burn fossil fuels, such as coal, crude oil or natural gas
- burn renewable **biomass**, such as manure, wood from trees or straw
- split **uranium** atoms into smaller ones.

Burning fuels turns the chemical energy of the fuel into heat energy. When the uranium atoms in uranium fuel rods are split, nuclear energy is changed into heat energy.

As well as being burned, biomass can also be fermented to produce methane, which can then be used as a fuel.

▲ We can generate electricity by splitting atoms

a Describe four different ways of heating water in power stations.

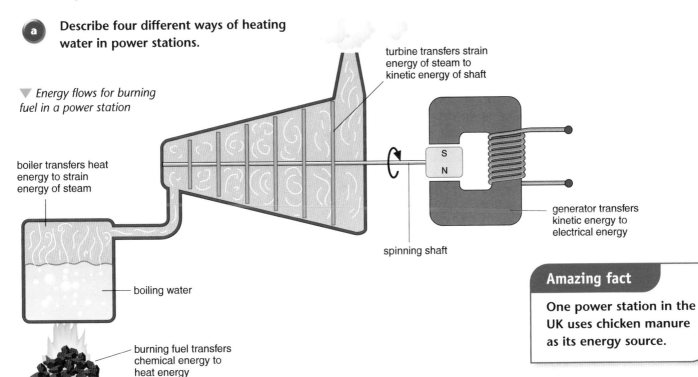

▼ Energy flows for burning fuel in a power station

turbine transfers strain energy of steam to kinetic energy of shaft

boiler transfers heat energy to strain energy of steam

spinning shaft

generator transfers kinetic energy to electrical energy

boiling water

burning fuel transfers chemical energy to heat energy

Amazing fact

One power station in the UK uses chicken manure as its energy source.

Which fuel?

The price you pay for your electricity depends on where you get it from. Electricity made from gas, coal and **nuclear power** stations is cheap. But they all produce polluting waste, which is bad for our environment.

The cost of sorting out the problems caused by these waste products in the future isn't included when you pay your electricity bill, but should it be? All of these cheap sources of energy are also limited and will eventually run out.

If you buy electricity from a renewable source, such as wind power, it is relatively expensive. Of course, you aren't paying for the fuel, because there isn't any. Instead, you are paying for the research and construction of the new technology.

But because wind power doesn't cause pollution there should be no need for future generations to pay the price of pollution.

Nuclear power

Nuclear power stations use uranium as their fuel. There are lots of advantages and disadvantages to using nuclear power.

Plutonium is a waste product left over after the uranium fuel has been processed. It is a radioactive material, so the ionising radiation from it can cause cancer. Plutonium is also a problem because it can be used to make nuclear bombs.

Advantages	Disadvantages
nuclear power uses advanced technology	uranium is a non-renewable fuel so stocks will eventually run out
at the moment there is plenty of uranium available	when a nuclear reactor is shut down the decommissioning costs are high
nuclear fuel does not produce greenhouse gases which may contribute to global warming	there is a risk of accidental emissions from radioactive materials
nuclear power does not rely on fossil fuels which are running out	nuclear power stations cost a lot to run

b What the advantages and disadvantages of using nuclear fuel for making electricity?

◄ *Radioactive materials need to be handled carefully*

Paying for electricity

Electricity meters record the amount of electrical energy that passes through them, in units called **kilowatt-hours** (kWh).

The cost of each kilowatt-hour (or unit) depends on the time of day and how the electricity is generated. It can be cheaper to buy electricity at night (but only if you have Economy 7).

So how much does it cost to run an electrical device? The cost depends on the **power** of the device and how long it is turned on for. First you need to calculate the power of the device from its current and voltage using this equation:

power = current × voltage

If a heater uses a current of 12 A (amps) and a voltage of 230 V (volts) then its power in **watts** is:

power = 12 A × 230 V = 2,760 W

There are 1,000 watts in a **kilowatt**. So to convert the power in watts (W) to kilowatts (kW) you need to divide by 1,000:

power = 2,760 W/1,000 W = 2.76 kW

c A 72 W laptop computer draws a current of 6 A from its battery. What is the battery voltage?

So how much does it cost to run a 3 kW heater for two hours when a unit of electricity costs 8 p? You can use this equation to calculate the energy supplied in kilowatt-hours:

energy supplied = power × time

energy supplied = 3 kW × 2 h = 6 kWh

The cost is then:

6 kWh × 8 p/kWh = 48 p

d A 100 W lamp is left on for 12 h. If a unit costs 9 p, what is the cost of running the lamp?

Examiner's tip

Show every step in your calculations.

keywords

biomass • kilowatt • kilowatt-hour • nuclear power • power • plutonium • uranium • watt

Green and clean

▲ *Cars can also be run on biodiesel*

In 2002 the supermarket chain ASDA seriously considered running its fleet of delivery trucks on fuel made from the waste fat and oil from its restaurants and rotisseries.

Each year it throws away 50 million litres of used cooking oil and fat. This can be converted by a simple chemical process into biodiesel, a fuel which can be used to run existing diesel engines. At about 10p per litre, it costs about the same to produce as dinodiesel, the diesel fuel derived from crude oil.

The tax levied on biodiesel by UK and European governments is about half that levied on dinodiesel, making it very attractive for people who run diesel engines for their business.

- Biodiesel is renewable because it comes from biomass.
- It doesn't contain sulfur impurities, unlike crude oil.
- As a product of biomass, it doesn't contribute to global warming.
- It is made from material which would otherwise be thrown down the drains or into landfill.

Biodiesel sounds like the ideal fuel for personal transport in the future. The government clearly thinks so, which is why it levies so much less tax on biodiesel.

But why do we still rely so heavily on dinodiesel? This is possibly because oil companies have too much money invested in extracting and processing crude oil to want to encourage us to switch from dinodiesel to biodiesel.

Biodiesel does not require a huge investment. The equipment to process 300 litres of vegetable oil each day has a payback time of less than a year. Apart from the fat or oil, you only need sodium hydroxide, water and methanol. The methanol is recycled and the only waste product is glycerine, which is a useful material for the chemical industry.

Questions

1 What is the difference between dinodiesel and biodiesel?

2 When biodiesel is burnt, it gives off carbon dioxide. Suggest why this does not lead to global warming.

3 Explain why large supermarket chains should consider using biodiesel.

4 Suggest three reasons why biodiesel has much less tax on it than dinodiesel.

5 Explain why large oil companies may not be keen to sell biodiesel.

Kill or cure?

In this item you will find out

- how nuclear radiation ionises materials
- about alpha, beta and gamma radiation
- how to deal with radioactive waste safely

Radioactive materials give out **nuclear radiation**. Radioactivity was only discovered about a hundred years ago. This is because you need special devices to detect the nuclear radiation. So until the invention of photographic film, nobody knew that radioactivity existed.

Radioactivity was discovered by Henri Becquerel in 1896. He found that rocks made from uranium emitted something which could trigger chemical reactions in photographic film wrapped in black paper. It was soon established by Marie Curie that only certain substances gave out these radiations.

Radioactivity has many uses – some of which have serious consequences for our future. Its use in medicine and industry provides enormous benefits. It can kill germs and cancer cells and it can be used to make electricity without contributing to global warming. But our understanding of radioactivity also gives us the technology to wipe our species off the planet.

You are surrounded by radiation all the time. Sources of this **background radiation** include:

- **cosmic rays**; fast moving particles from outer space
- radon; a radioactive gas given off by soil and rocks in the ground
- your food and drink and other living things
- radioactive substances.

There is no way of escaping background radiation, but the risk it poses to your health is quite small.

▲ Henri Becquerel in his laboratory

▲ Testing livestock from some parts of the UK reveals they are still contaminated with radioactive fallout from the explosion of the Chernobyl reactor in 1986

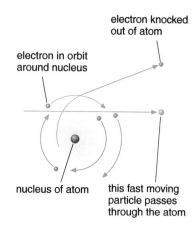

electron knocked out of atom

electron in orbit around nucleus

nucleus of atom

this fast moving particle passes through the atom

Nuclear radiation

When nuclear radiation passes through materials it ionises them. This means that radiation (fast moving particles that carry a lot of energy) knocks electrons out of particles as it passes through a material and causes other particles to gain electrons. This produces charged particles. The **ionisation** of particles can result in chemical changes which are bad for living cells.

a Explain how nuclear radiation produces ionisation.

Types of nuclear radiation

There are three types of nuclear radiation:

- **alpha** particles, which can be stopped by a sheet of paper
- **beta** particles, which need a few millimetres of metal to be stopped
- **gamma** rays, which are partially absorbed by a few centimetres of metal.

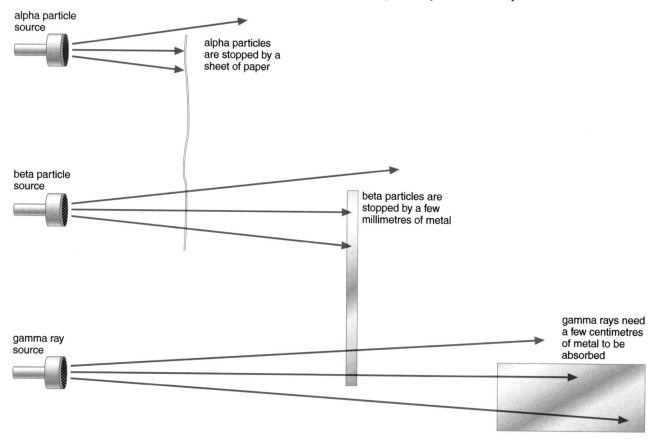

alpha particle source

alpha particles are stopped by a sheet of paper

beta particle source

beta particles are stopped by a few millimetres of metal

gamma ray source

gamma rays need a few centimetres of metal to be absorbed

b Explain why alpha particles are the least dangerous, for humans, of the three nuclear radiations outside the body.

Radiation at work

All three types of radiation are useful. Smoke detectors contain a small amount of americium-241, an emitter of alpha particles. Any smoke alters the ionisation of the air caused by the alpha particles which triggers the alarm.

Thickness gauges for the continuous production of sheets of paper or plastic use beta particles from strontium-90. Beta particles are fired at the sheet and a detector is placed on the other side. As the thickness of the sheet increases, the number of beta particles detected decreases.

Gamma rays can be used in a similar way to investigate the internal structure of metal welds. The gamma ray source is placed on one side of the weld and a sheet of photographic film on the other. As the amount of metal in the way increases, the number of gamma rays getting through to the film decreases.

Beams of gamma rays from cobalt-60 are used to kill tumours inside people, avoiding the need for surgery. Medical equipment, such as scalpels and bandages, are sterilised by gamma rays. Items are wrapped in plastic and left close to some cobalt-60 for several hours. The gamma rays penetrate deep inside the item, killing any living organism which might cause infection.

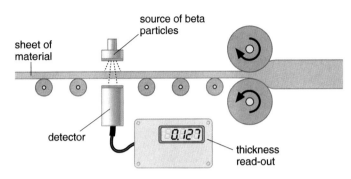

▲ *Measuring thickness with beta particles*

c Describe one use for each of the three nuclear radiations.

d With the help of some examples, describe the properties of gamma rays which make them useful in medicine.

Radioactive waste

There is also a problem with using radioactive materials. What do you do with the **radioactive waste** when you have finished with it? You cannot just leave it lying around, because its radiations are harmful. Some nuclear wastes contain low levels of radioactivity and can be buried safely in trenches in the ground.

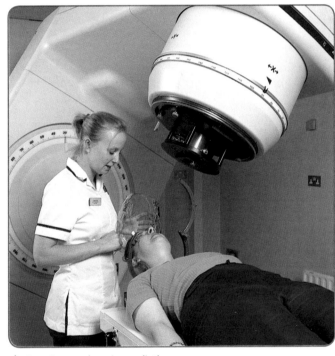

▲ *A patient undergoing radiotherapy*

However, some of the waste from nuclear power stations will remain very radioactive for many thousands of years. At the moment there are three possible ways of dealing with this problem:

- Bury the waste in landfills. You can only do this if the radioactivity is low enough.
- Reprocess the waste. Some of the material in the waste can be used to make fresh nuclear fuel.
- Encase the most dangerous waste in glass blocks which can be stored safely underground.

There is a risk with burying the waste that radioactive materials might leak into the groundwater and contaminate it. Also, acceptable levels of radioactivity may change in the future. If they became less, then we would have even more of a problem disposing of the waste.

keywords

alpha • background radiation • beta • cosmic rays • gamma • ionisation • nuclear radiation • radioactive materials • radioactive waste

The hottest problem of our time?

Nuclear power stations have a finite lifetime, during which their core becomes radioactive. Nobody has yet managed completely to dismantle a decommissioned nuclear power station, let alone safely dispose of its highly radioactive contents.

So what happens to the radioactive waste? You can't leave it lying around, because its radioactivity is harmful. It needs to be guarded, so that the plutonium is kept out of the hands of terrorists who could use it to make bombs.

At present, waste from nuclear power stations is treated as follows:

• It is left in ponds of water to let some of its radioactivity die down. The water heats up as it absorbs the radiation, so it needs to be cooled as well as guarded.
• It is then processed to separate the plutonium and uranium from the rest. This is expensive as it has to be done by automatic machinery that can survive the lethal levels of radioactivity.
• Material which is only slightly radioactive can then be buried in landfill.
• The rest of the highly radioactive waste is stored in metal drums on secure sites until the government decides how to dispose of it.

▲ *Nuclear reactors are housed in these huge concrete buildings*

Questions

1 What happens, at present, to a nuclear power station at the end of its useful life? Why is this a problem?

2 Explain why the waste is stored under water before it is reprocessed.

3 Suggest why nuclear waste is reprocessed.

4 Suggest why governments have yet to decide on the final resting place of highly radioactive waste.

The cosmic ray shield

In this item you will find out

- about the Earth's magnetic field

- how the Moon may have been created

- about solar flares

Three thousand years ago, the Chinese invented the magnetic compass. They floated an iron-rich rock called magnetite on water and found that it always pointed in the same direction.

We now know that a magnetic compass indicates the direction of the Earth's **magnetic field**.

It is only in the last 50 years that we have understood that this field shields us from the worst effects of cosmic rays – deadly fast-moving particles from the Sun and outer space.

Without this shield, the molecules that made up life in the beginning might never have survived to evolve into humans.

But where does the Earth's magnetic field come from? And why is it so strong? The magnetic field around similar-sized planets is much weaker.

▲ The compass is a crucial aid to navigation on land or at sea

One theory involves the collision of two large planets. All of the heavy iron ends up in the larger planet, allowing it to have a large magnetic field and be safe for living organisms. Most of the rest of the lighter material ends up in orbit as a large moon with no magnetism at all.

A lot of evidence supports this as the origin of our own Earth–Moon system. Perhaps life on Earth is only possible because of a catastrophic collision a long time ago.

 Suggest an explanation for the large amount of iron at the centre of the Earth.

Amazing fact

The Earth's magnetic field has reversed its direction 171 times in the last 76 million years.

Cosmic rays

The Earth is battered on all sides by cosmic rays. As these particles hit the upper layers of our atmosphere, they create gamma rays. These carry the energy of the particles down to ground level, where they can ionise materials. This is bad news for living organisms, as a single cosmic ray can create hundreds of gamma rays which can cause cancer.

Cosmic rays that are charged particles change direction as they pass through a magnetic field. The Earth's magnetic field reaches far beyond its atmosphere, so many cosmic rays spiral round its lines of force before entering the Earth close to the North or South Poles.

The dramatic display of colour, the Aurora Borealis, in the night sky near the poles is caused by cosmic rays. As they enter the atmosphere, they ionise the atoms. The light is created when the ions recombine with electrons to become neutral atoms once more.

▲ *The Aurora Borealis is caused by cosmic rays*

 What are cosmic rays? Why are they dangerous if they reach Earth?

Earth's magnetic field

The diagram shows the shape of the Earth's magnetic field. Like a magnet, the magnetic field has a north pole and a south pole.

Where does the Earth's field come from? Scientists think that it is made by large electric currents set up in the liquid core of iron as the Earth spins on its axis. The shape of the field is very like the one around a coil of wire when it carries a current. An electric current is just a flow of charged particles (electrons) through a metal. Magnetic fields are generated whenever charged particles move around.

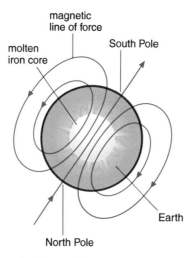
▲ *The Earth's magnetic field*

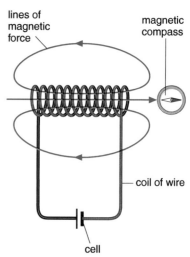
▲ *The lines of force round a current-carrying coil*

c Sketch the shape of the Earth's magnetic field. Explain how it protects life on Earth.

Making the Moon

Our planet has a very large molten iron core compared with other objects visited by our spacecraft. Without that iron, we would have no magnetic field.

The Moon does not have an iron core. Scientists think that a catastrophic event which may have happened four billion years ago could explain this.

A planet the size of Mars may have collided with the Earth. All of the heavy iron from the cores of both planets merged to create the Earth, while all the lighter material merged to form the Moon.

Astronauts have brought rocks back from the Moon. They are very similar to rocks on the surface of the Earth, suggesting that both do indeed have the same source.

d Explain one theory why the Earth has such a large magnetic field.

Solar flares

Solar flares are vast clouds of charged particles that are ejected from the Sun from time to time. The particles are ejected at high speed. They produce strong disturbed magnetic fields.

If solar flares get close enough to Earth they have enough energy to get inside the electronic circuitry of a communications **satellite** and permanently destroy its electronics.

e Why can solar flares interfere with our communications systems?

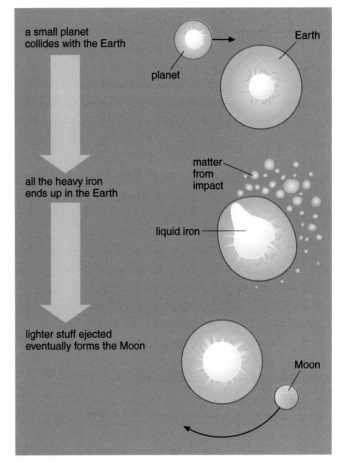

a small planet collides with the Earth

planet

Earth

all the heavy iron ends up in the Earth

matter from impact

liquid iron

lighter stuff ejected eventually forms the Moon

Moon

▲ How the Moon may have been formed

▲ This satellite will be used by scientists to observe solar flares

▲ A solar flare

keywords

magnetic field • satellite • solar flares

201

Power cuts from space

▲ *A coronal mass ejection from the Sun*

On March 13th 1989, Quebec's power supply failed completely, leaving millions of people without electricity for days.

The damage was caused by the Sun. A huge cloud of swirling charged particles had been launched in the Earth's direction the previous day, travelling at over a million miles per hour. It was a type of solar flare called a coronal mass ejection (CME).

When it slammed into the Earth's magnetic field over Canada, it was like two giant magnets colliding. The resulting surges of magnetism on the Earth's surface created pulses of electricity in large metal loops, such as those formed by transmission line systems.

There was enough energy in those pulses to burn out transformers, bringing the electricity distribution system to a sudden stop.

Next time, we may have advance warning. There are now satellites in orbit around the Sun, looking for CMEs heading towards Earth.

When one is detected, power distribution networks can be disconnected until the surge of magnetism is over. Without a complete circuit, the change of magnetism won't be able to produce any current to burn out transformers.

However, satellites and astronauts in orbit around the Earth may not be so lucky. Being hit by a CME is like being exposed to a lot of radioactivity at once. Electronic circuits can be designed to survive this, but shielding is the only answer for humans. And putting thick lead shielding in a spacecraft makes it very heavy and expensive to launch, so the only solution may be to get astronauts back to Earth before the CME arrives.

Questions

1 What is a CME?

2 Explain why a CME can cause a power failure on Earth.

3 Suggest why suddenly breaking a transmission line can damage a power station – where does the electricity go?

4 Why are CMEs dangerous for astronauts in orbit around the Earth? What can be done to protect them?

5 Satellites in Earth orbit can be damaged by CMEs. What effects could this have on Earth?

Is anybody out there?

In this item you will find out

- about the objects in our Solar System and their relative positions

- about the difficulties of manned space travel between planets

- how unmanned spacecraft allow us to explore the Solar System

Humans have always been interested in the stars and have watched for intelligent communication from them. We watched with naked eyes at first, then with optical telescopes and finally across the entire electromagnetic spectrum for messages from aliens.

As yet, there is no scientific evidence of anything out there trying to communicate with us. But we keep on listening.

Of course, life may be everywhere in the Universe, but intelligence capable of beaming signals across the galaxy may be rare. Our own species, *Homo sapiens*, has been around for about four million years, but we have only acquired the technology to beam out radio signals in the last 40 years.

Unmanned spacecraft are visiting the most likely habitats for life in the Solar System and the evidence is beginning to suggest there might be bacteria on other planets.

For example, there are traces of methane in the atmosphere of Mars. This is destroyed by ultraviolet radiation from the Sun, so something must be replacing it as fast as it is destroyed. Bacteria can generate methane on Earth, so perhaps they are doing it on Mars.

▲ *This robot spacecraft crashed on Mars. It was equipped to detect the presence of life in the soil*

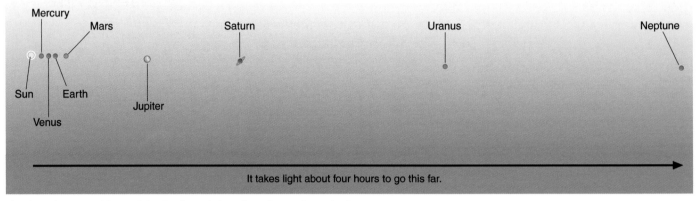

It takes light about four hours to go this far.

▲ *The relative positions of the Earth and the other planets from the Sun*

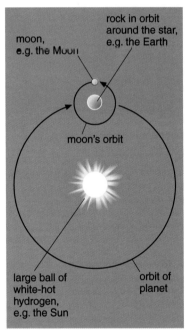

▲ *The Moon orbits the Earth which orbits the Sun*

Our Solar System

We inhabit the surface of a **planet**, one of nine in **orbit** around a large ball of white-hot hydrogen gas. This is the **star** that we call the Sun. Like all stars, it is so hot that it emits huge quantities of light. There is a lump of rock which orbits around our planet. We call it the Moon. The Earth has only one moon, but other planets have many more. Some have none.

Round and round

Most of the planets follow near-circular paths, centred on the Sun. Any motion in a circle requires a **centripetal force** towards the centre of the circle.

For each planet, its centripetal force is supplied by the Sun's gravitational force. In the absence of this force, each planet would carry on moving in a straight line. As each planet moves forwards, gravity tugs at right angles, changing the direction of motion. The speed of each planet is just right for this to result in stable near-circular orbits for most of the planets.

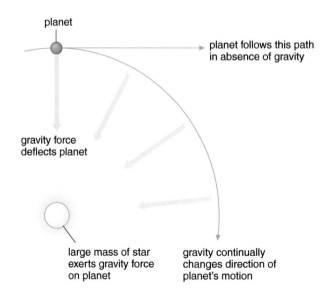

a **Explain how satellites are kept in circular orbits around the Earth.**

Space travel

Getting humans to other planets is going to be difficult, as they will need to cross vast distances of space. A **light-year** is a measurement of a very large distance and is the distance that light travels in one year.

Any voyage will take a very long time, perhaps many years. A **spacecraft** will need to have enough fuel for the journey and the astronauts will need to be provided with enough food and water.

The spacecraft will need to be designed so it provides the astronauts with a warm, stable atmosphere with plenty of oxygen. It will also need to have shielding to protect the astronauts from cosmic rays, which will increase the risk of cancer.

One major problem is that humans have evolved on Earth to survive in a gravity of strength 10 N/kg. They suffer from severe health problems if they are exposed to zero gravity for long periods of time.

 b List the problems of space travel for humans.

Amazing fact

Some of the first robot spacecraft have already left the Solar System. They won't arrive at the nearest stars for thousands of years.

Robots in space

Unmanned spacecraft, such as probes, have done most of the exploration of our Solar System. Onboard computers control the spacecraft, following commands sent by radio from Earth.

Computers don't need food, air or water, so robot spacecraft can easily be launched with today's rockets. They can also explore areas where conditions are lethal to humans and they do not suffer from the effects of zero gravity.

Their cameras have given us stunning images of other planets, using infrared and ultraviolet as well as visible light. Other sensors on the spacecraft can send back information about the temperature, magnetism, gravity and atmosphere on other planets.

There is no need for unmanned spacecraft to make the journey back to Earth and dozens of robot missions can be sent for the same cost as a single manned mission, so it doesn't matter if a few fail.

Two problems with unmanned spacecraft are reliability and maintenance. When the spacecraft are miles from the Earth there is no-one around to carry out repairs.

c What are the advantages of exploring the Solar System with robots instead of humans?

d Suggest one advantage of a manned mission to a remote planet instead of an unmanned one.

▲ Life in space

keywords

centripetal force • light-year • orbit • planet • spacecraft • star

Problems with gravity

Some of the hazards of space travel are obvious. The life support systems may fail, the rockets which fling you into space may blow up, cosmic rays may tear through you and give you cancer. But the biggest threat is the damage your body inflicts on itself.

Life at low gravity may sound fun, but it has serious consequences for your body.

Your bones are no longer stressed by the running, jumping and walking of normal life. Your body recognises this and doesn't waste resources on repairing them. This means that you lose at least 2% of your bone mass per month. At that rate, you lose half of your bone mass in three years.

Your muscles shrivel as your body finds that you are no longer using them as much as before. As your body adapts to low gravity, you could lose 20% of your muscle tissue.

Your heart becomes weaker as it no longer has to pump blood from your feet to your head against gravity.

The first astronauts to visit Mars will take over a year to get there. Gravity on Mars is less than on Earth, but they may be unable to walk on the surface of the planet without breaking a limb. By the time they get back to Earth, none of them will be able to support their own weight.

So what can be done about this? One solution may be to set the spacecraft spinning. Astronauts inside the spacecraft will need a force from the hull to keep them moving in a circle. Get the speed right and this force will mimic Earth's gravity.

▲ *Warning: space travel can seriously damage your health*

hull pushes on astronaut to keep her moving in a circle

astronaut in artificial gravity

spacecraft spins around

outer hull of spacecraft

Questions

1 List some hazards of space travel.

2 Describe and explain the effects of zero gravity on the human body.

3 Explain how spinning the spacecraft could provide artificial gravity.

4 Once the spacecraft is spinning, no fuel will be needed to keep it going. Explain why.

Catastrophe!

In this item you will find out

- the difference between an asteroid and a comet

- about the evidence for asteroid collisions

- about Near-Earth Object observations

If you look at the Moon when it is full, it looks perfectly smooth and round. If you look at the Moon through a telescope you will notice the roughness of the surface, with its flat plains ringed by tall mountains.

Each of the circular features on the Moon's surface looks like an impact **crater**. This is what you get when a small object hits a much larger one with enough energy to melt part of it. The impact pushes hot plastic rock into a ring of material, surrounding a flat plain where the molten rock solidifies again.

The Moon is covered with many craters – large ones, small ones and overlapping ones. Scientists use this evidence of violent impacts to support their ideas of planet formation.

Gravity tugs dust and rocks towards each other, so that they collide and stick. As objects grow in this fashion, the impacts get more and more spectacular, until all the loose matter has been swallowed up into a moon or a planet. With no wind, frost or rain to change its surface, the record of those violent times is preserved forever on the surface of the Moon.

But there are craters on Earth which look just like those on the Moon. They can't be old, or wind and rain would have worn them away by now. So when did these impacts happen and will they happen again in the future?

a Explain why the Moon is covered in craters, but the Earth is not.

WHY THE MOON IS COVERED IN CRATERS

▲ *The moon is covered in craters like these*

Asteroids

Asteroids are large rocks left over from the creation of the Solar System. Lots of asteroids are orbiting in the Asteroid Belt between Mars and Jupiter. The belt is between Mars and Jupiter because the large gravity of Jupiter disrupts the formation of planets so the rocks cannot join together.

Collisions can push asteroids out of a safe circular orbit, well away from the Earth, into an elliptical orbit which crosses Earth's orbit.

 Why are asteroids safe in circular orbits but unsafe in elliptical ones?

▲ An impact crater on Earth

▲ The Asteroid Belt

Past collisions

There is evidence that asteroids have collided with the Earth in the past. Impact craters on Earth provide some of this evidence.

Scientists think that the last collision was 65 million years ago, which may have caused the extinction of the dinosaurs.

The evidence for this last big collision is to be found in sedimentary rocks deep underground. One layer of sedimentary rocks, found over the entire surface of the Earth, contains lots of iridium. This is a metal element which is rare on Earth and may have come from the impact object itself.

An asteroid collision would cause widespread fires on Earth and this layer of rock also contains lots of carbon which would have been produced by the fires.

There are lots of fossils in the layer of rock underneath, which suggests that life was widespread before the collision, but very few fossils in the rock layer above, which suggests that life after the impact was sparse.

Also, seismic exploration of the rocks under the Gulf of Mexico has suggested the presence of a crater under sedimentary rocks laid down less than 65 million years ago.

c What is the evidence for an impact with an object from space 65 million years ago?

Examiner's tip

Always include evidence in your explanation.

Looking out for NEOs

A **Near-Earth Object** (NEO) is anything big in space, such as an asteroid or a **comet**, which might collide with the Earth.

Both could cause a lot of damage if they hit us. Astronomers track them with telescopes and calculate the chances of a collision in the future. But the relatively small mass of an NEO means that its orbit is easily changed by other planets it passes.

As well as tracking Near-Earth Objects with telescopes, scientists could also use satellites to monitor them in order to reduce the threat of a collision with a Near-Earth Object.

If one looked like it might hit the Earth, it might be possible to deflect it from its path by an explosion.

d What could we do to avoid being hit by an NEO?

Comets

Comets are made from material orbiting the Sun from beyond the planets. They are cold chunks of ice and dust in large slow orbits beyond Pluto. They can be jolted into **elliptical** orbits around the Sun or other stars by the gravity of passing stars.

A comet spends most of its time far away from the Sun. As it approaches the Sun it speeds up in the stronger gravity. Heat from the Sun melts the ice, so it sheds material to form a tail.

e Comets don't last for ever. Why not?

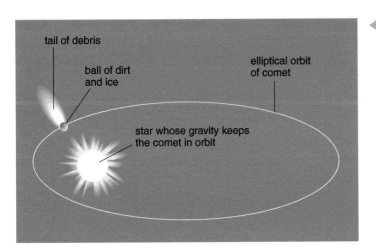

◀ *The orbit of a comet*

tail of debris

ball of dirt and ice

elliptical orbit of comet

star whose gravity keeps the comet in orbit

▲ *Halley's comet*

keywords

asteroid • comet • crater • elliptical • Near-Earth Object

A near miss for Earth

Herald Times Friday June 21 2002
Science Correspondent

An asteroid the size of a large supermarket travelling at 24,000 mph just missed the Earth last Friday. Nobody noticed it at the time.

In fact, it was not spotted until three days later by David Iron, an astronomer.

'Optical telescopes can't see anything on the sunward side of Earth,' says David, 'At this time of year, Earth is hit by lots of small asteroids. This one was just a lot bigger.'

David has calculated that the asteroid, called 2002MN, missed Earth by a mere 75,000 miles. This sounds a lot, but it is about one-quarter of the distance from Earth to Moon.

'The asteroid is about 100 m across,' says David, 'with an impact energy of a thousand nuclear bombs. If it had hit England, the loss of life would have been huge.'

Something with this amount of energy fell over Siberia in 1908. It flattened huge areas of forest, but few people were around to see it happen. The last time that a large asteroid passed between Earth and Moon was in December 1994, according to the Near-Earth Object (NEO) Information Centre based in Leicester. So when is the next one due? And will it miss?

Nobody really knows. NASA uses telescopes in the northern hemisphere to keep track of asteroids which are bigger than 1 km across, but there is no government money for tracking smaller objects. There is no programme searching for NEOs approaching the southern hemisphere, whatever their size.

▶ *This may have been caused by an asteroid*

Questions

1 Explain how an object only 100 m across could damage an area over 1,000,000 m across.

2 Suggest why nobody from the southern hemisphere is looking for NEOs.

3 What do you think NASA should do when it finds an object over 1 km across on a collision course with Earth?

Life and death of the stars

In this item you will find out

- about the Big Bang theory of the Universe

- the life history of a star

- what happens when stars die

Our knowledge of the Universe comes from interpreting its electromagnetic waves. Telescopes allow us to survey the sky over the whole electromagnetic spectrum, from radio waves to gamma rays.

All the waves seem to tell us one story – that the Universe appeared explosively out of nothing 15 billion years ago and that it is still expanding today.

The key to extracting this story from starlight was discovered by accident. In 1814 Joseph von Fraunhofer noticed that when a glass prism split sunlight into colours, there were black bands in the spectrum.

In 1823 he tried the same experiment with starlight and discovered that the black bands were in different places. They were shifted to the red end of the spectrum and they had a longer wavelength.

In 1842 Christian Doppler put the last piece in the jigsaw by discovering that the wavelength of a wave increases as you move away from its source. So, stars with red-shifted light must be moving away from us. The red shift of starlight means that the Universe is expanding!

But will it expand for ever? It all depends on the mass of the Universe. If it contains enough matter, then gravity will eventually pull everything back into a final crunch. If there isn't enough mass, then the Universe will go on expanding for ever.

▲ *Fraunhofer noticed black bands in the Sun's spectrum*

▲ *Starfield from Hubble telescope*

a **The light from the star Glog is shifted towards the blue end of the spectrum. What does this tell you about Glog?**

The expanding Universe

Stars are clumped together in galaxies, with about a billion stars in each. Scientists can measure the speed of these galaxies by looking at the red shift of the light from their stars. These measurements provide evidence that:

- Almost every galaxy is moving away from ours.
- The further away a galaxy is, the faster it moves.

▲ *A galaxy like our Milky Way*

There is no reason to suppose that our own galaxy, the Milky Way, is at the centre of the Universe. The only way of explaining the fact that, whichever direction you look in space, almost everything else appears to be rushing away from you, is to assume that space itself is expanding!

 Describe the evidence for an expanding Universe.

The Big Bang theory

The **Big Bang theory** explains how the Universe may have developed:

- A very hot Universe appeared 15 billion years ago, at a single point.
- It expanded rapidly and cooled at the same time.
- Atoms (mostly hydrogen) formed when the temperature dropped to about 3,000 °C.
- Gravity clumped hydrogen atoms into galaxies and stars.

The Big Bang theory accounts for why the light from galaxies is red shifted and why the further away the galaxy is, the greater is the red shift. It also explains the possible starting point of the Universe and its age.

The Universe is filled with microwave radiation. This is because the radiation left over from shortly after the Big Bang has been stretched from very short wavelength gamma radiation to centimetre-long microwave radiation as the Universe expanded.

 Describe the Big Bang theory.

Making stars

The Universe is full of interstellar gas clouds full of hydrogen gas. How do they become stars?

Gravity in the gas clouds tugs every atom towards every other atom so the cloud experiences gravitational collapse and shrinks. It becomes a proto-star.

The gravitational energy is transferred to heat energy, until thermonuclear fusion reactions start at the hot centre of the proto-star. The fusion reactions start at a temperature of about 15 million °C and convert hydrogen into helium. This releases energy and stops the star shrinking any more. The star then has a long period of normal life.

 d   **Explain how a cloud of hydrogen becomes a star.**

Star death

The end of a star depends on the mass of the star. What happens when a medium-weight star, like our Sun, starts to run out of hydrogen fuel? It becomes unstable, swelling up to form a **red giant** as it converts helium into carbon. It will then cool down and shrink to form a **white dwarf**.

But do not worry – the Sun has enough fuel for another 5 billion years. Before a star dies it ejects lots of gas and dust. This is called **planetary nebula** and it is what new stars are made from.

If the star is a heavy-weight star, it will first swell into a red giant before it explodes as a **supernova**.

No star can use fusion to form elements with atomic mass above that of iron. All of the heavier elements are created in a few seconds, as a supernova tears a star apart. These elements are scattered through the surrounding clouds of hydrogen, ready for gravity to make them collapse once more into new stars and planetary systems.

If the star is massive enough, the explosion will leave a **neutron star** or a **black hole**. A black hole is an object where its mass is so large and its gravity is so strong that not even light can escape from it.

 e   **Describe what happens to a heavy-weight star when it runs out of hydrogen fuel.**

Amazing fact

The black hole at the centre of our galaxy probably swallows a star each year. It is about the size of our Solar System, but is a million times heavier.

keywords

Big Bang theory • black hole • neutron star • planetary nebula • red giant • supernova • white dwarf

The age of time itself

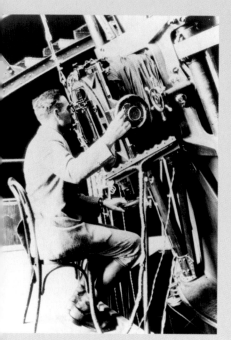

▲ *Edwin Hubble at work*

If the Universe is expanding, then at some time in the past it must have all been at one point. When was that? Edwin Hubble, who published the first evidence for an expanding Universe in 1929, reckoned that it was 2 billion years ago.

As you can see from the graph below, Hubble compared the speed of just 24 galaxies with their distance. He drew a best straight line through the points, concluding that galaxies which are about 3.3 million light years apart today are moving apart from each other by 500 km/s. If that speed has been constant since the beginning of time, then it is a simple matter to calculate how long ago that was:

light moves at 3×10^5 km/s

a year contains $365 \times 24 \times 60 \times 60 = 3.2 \times 10^7$ s

so a light year is 3×10^5 km/s $\times 3.2 \times 10^7$ s $= 9.5 \times 10^{12}$ km

3.3 million light years is therefore 9.5×10^{12} km $\times 3.3 \times 10^6 = 3.1 \times 10^{19}$ km

time = distance/speed = 3.1×10^{19} km/500 km/s $= 6.3 \times 10^{16}$ s.

We can work out the speed of a galaxy but how can we tell how far away it is? Hubble used the most advanced telescope of his day to show that galaxies contain Cepheid variables – stars whose brightness oscillates with time.

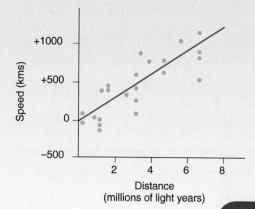

There are similar Cepheids in our own galaxy that are close enough for us to measure their distance. A star that is ten times further away will appear a hundred times fainter, so the distance to another galaxy can be estimated by comparing the brightness of its Cepheids to our local ones.

Modern measurements of the distance to local Cepheids suggests that the whole Universe came into being 15 billion years ago. Without a Universe, there is nothing to make a clock with, so the age of the Universe is the age of time itself!

Questions

1 Show that 6.3×10^{16} s is about 2 billion years.

2 How do we know how fast other galaxies are moving away from us?

3 How do we know how far away other galaxies are?

4 It takes 17 minutes for a pulse of microwaves from the Earth to return after reflecting off the Sun. What is the distance, in kilometres, between Earth and the Sun?

5 Modern measurements suggest that galaxies which are one million light years apart are moving away from each other at 20 km/s. Use this to calculate the age of the Universe.

P2a

1 Photocells can be used to make electricity.

State two advantages and two disadvantages of using photocells. [4]

2 Describe three different ways in which energy from the Sun can be used to provide energy for our homes. [3]

3 Use ideas of electrons to explain how light energy is transferred to electricity in a photocell. [3]

4 The sentences explain how passive solar heating works. Put them in the correct order:

A raising the temperature of the air between wall and glass
B the wall transfers light energy to heat energy
C infrared radiation is reflected by the glass
D light energy is transmitted through glass [3]

5 Describe two advantages and two disadvantages of using wind turbines to generate electricity. [4]

6 Some people think that wind turbines are best placed out at sea.

Suggest what are the advantages and disadvantages of this. [4]

7 A solar heater has water pumped through pipes exposed to the Sun. The pipes are kept under glass and coloured black. The sentences explain how a solar heater works. Put them in the correct order.

A Sunlight passes through glass.
B The infrared is trapped by the glass.
C The pipes heat up emitting infrared radiation.
D The water carries the heat energy into the house.
E Light energy is absorbed by the surface of the pipes.
F Heat energy conducts through the pipes to the water. [5]

P2b

1 A generator has a coil of wire next to a spinning magnet.

State three ways of increasing the voltage from the generator. [3]

2 The sentences describe the energy transfers in a power station. Put them in the correct order:

A which makes the generator spin round
B which passes through the turbine
C water is boiled to make steam
D the fuel is burnt [3]

3 Explain how transformers are used in the national grid. [4]

4 A wood-burning power station has an efficiency of 0.35.

a For every 500 J of chemical energy transferred into the power station, how much electrical energy is transferred out? [3]

b How much waste heat energy is transferred out for each 500 J transferred in? [1]

5 Explain why increasing the voltage of the National Grid saves energy. [2]

6 Power stations produce a lot of waste heat, in the form of water at about 90°C.

Suggest two ways of putting that heat to use, and suggest why it is not already happening in the UK. [4]

7 The sentences explain how electricity is moved throught the national grid. Put them in order.

A Electricity leaves the power station.
B Its voltage is lowered by a transformer.
C Its voltage is increased by a transformer.
D The electricity is delivered to the customer.
E The electricity passes along the transmission lines. [4]

P2c

1 A kettle has a current of 9 A when connected to a 230 V supply.

Use the equation power = voltage × current to calculate the power of the kettle. [2]

2 Calculate the cost of running a 2 kW heater for 6 hours when a unit of electricity costs 8p. [3]

3 Match the start and end of the sentences:

1 Methane is a product   a of a nuclear reactor

2 Plutonium is a waste product   b of burning a fuel.

3 Heat energy is transferred as a result   c of fermenting biomass [2]

4 The cost of buying electricity to run a computer continuously for a week is £1.68.

What is the power of the computer? The cost of a unit of electricity is 10p. [3]

5 Describe two advantages and two disadvantages of using nuclear power to generate electricity. [4]

6 A power station can burn coal or wood to generate electricity.

Describe two advantages and two disadvantages of using wood instead of coal. [4]

7 A power station generates electricity by burning sawdust from a sawmill.

Explain why the waste gases from this power station do not contribute to global warming, even though they are made from carbon dioxide. [2]

8 The sentences explain how electricity can be made from rubbish. Put them in the correct order.

A Burn the gas in a turbine.
B Dig a big hole in the ground.
C Fill the hole with wet rubbish.
D Line it with a gasproof material.
E Cover the hole with a gasproof top.
F Make electricity by spinning a generator.
G Collect methane gas from rotting rubbish. [6]

P2d

1 What is background radiation? Where does it come from? [2]

2 Describe one use each for alpha particles, beta particles and gamma rays. [3]

3 Describe two ways of disposing of radioactive materials. [2]

4 Nuclear radiation ionises materials. Explain what this means. [2]

5 Describe and explain two problems associated with the disposal of radioactive waste from nuclear power stations. [4]

6 You are provided with a radioactive substance.

Describe how you would determine the type of radiation emitted by it. Include a description of the apparatus required. [4]

7 **a** Describe how to dispose safely of waste materials that are highly radioactive. [2]

   b Why are these substances such a problem? [1]

8 Here are some ways in which radioactivity is used. For each use, explain what type of radiation (alpha, beta or gamma) must be used.

   a Radioactive material injected into a person can be used to find out where their blood flows. [2]

   b The amount of radiation passing through a material can be used to measure its thickness. [2]

   c Exposure to radiation can destroy a cancerous tumour. [2]

   d Smoke can stop radiation being detected, setting off a fire alarm. [2]

P2e

1 Describe how a collision between two planets can give rise to an Earth—Moon system. [3]

2 The Earth's magnetic field protects us from solar flares.

   a What is a solar flare? [2]

   b How does the magnetic field protect us? [1]

3 Explain how cosmic rays can cause the Aurora Borealis. [3]

4 What is the evidence for the creation of the Earth–Moon system from the collision of two planets? [3]

5 Solar flares sometimes hit the Earth. Describe two consequences of this. [2]

6 Complete the sentences.

Cosmic rays are mostly ____(1) particles from the ____(2).

They are deflected by the ____(3) field around the Earth, so not many of them get to its ____(4) where they can damage living ____(5) .

This protection does not exist at the North and South ____(6) . [6]

7 Here are some statements about the Moon. Which of them are evidence to support the idea that the Moon was created from the collision between a planet and the Earth?

A There is no air on the Moon.
B The Moon has no magnetic field.
C The Earth has a large magnetic field.
D The moon is smaller than the Pacific Ocean.
E The surface of the Moon is covered in craters.
F Moon rocks brought back by astronauts are similar to rocks on the Earth's surface. [3]

P2f

1 Providing enough food and water will be difficult for a manned voyage between planets.

State four other difficulties. [4]

2 Name the third and fifth planets from the Sun. [2]

3 Unmanned spacecraft have sent back information about other planets.

State four different things they have told us. [4]

4 The following objects make up the Solar System: planets, star, comets and asteroids.

Describe their place in the Solar System. [4]

5 The nearest star is 4 light-years away from Earth. What does this mean? [2]

6 Unmanned spacecraft do not need a supply of food, air or water.

Describe three other advantages of using unmanned spacecraft to explore the Solar System. [3]

7 The sentences explain how a planet's distance from the Sun affects its temperature. Put them in the correct order.

A They spread out in straight lines as they travel through space.
B A far off planet only intercepts a few of these rays.
C A nearby planet intercepts lots of these rays.
D So it absorbs lots of energy from the Sun.
E So it absorbs less energy from the Sun.
F Light rays are emitted from the Sun. [5]

8 Match the start and end of the sentences.

1 A moon is   a   a glowing ball of hydrogen gas

2 A galaxy is   b   a large lump of rock which orbits a star

3 A meteor is   c   a small lump of rock which orbits a star

4 A planet is   d   made from billions of stars

5 A star is   e   a lump of rock which orbits a planet. [4]

P2g

1 There is evidence on Earth for collisions with asteroids in the past. Describe three pieces of the evidence. [3]

2 A comet recently hit Jupiter.

a Where do comets come from? [1]
b What are comets made of? [1]
c Why does a comet have a tail? [2]

3 How do scientists know if a NEO is going to impact on Earth? [2]

4 Explain why the majority of asteroids are in orbit around the Sun between Mars and Jupiter. [2]

5 Comets have a highly elliptical orbit around the Sun.

Describe and explain how the speed of the comet changes as it goes round its orbit. [4]

6 Describe and explain the actions which could be taken to prevent a NEO from impacting on the Earth. [3]

7 Plants in a spaceship could be useful for long voyages in space. Match the start and end of the sentences.

1 The air breathed out by the astronauts   a   provides energy for the plants

2 The oxygen given out by the plants   b   provide food for the astronauts

3 The roots and seeds of the plants   c   provide fertiliser for the plants

4 The astronauts' waste products   d   is breathed in by the astronauts

5 Light from the Sun   e   is made into food by the plants [4]

P2h

1 The sentences are about the story of the Sun. Put them in the correct order:

A a cloud of hydrogen gas collapses under its own gravity
B heating up until fusion reactions start to convert hydrogen to helium
C stopping the collapse by emitting heat and light
D when hydrogen runs out the star swells to a red giant
E then collapses and cools to form a white dwarf [4]

2 Describe the end of a large star. [3]

3 How is the speed of a galaxy related to its distance? [3]

4 Describe and explain the difference between a white dwarf and a black hole. [4]

5 Explain how the red shift of light from galaxies provides evidence for the Big Bang theory. [4]

6 What two things does the red shift of the light from a galaxy tell us about that galaxy? [2]

7 Match the start and end of the sentences.

1 A red giant is   a   a star which has stopped fusion reactions

2 A planetary nebula is   b   an exploding star

3 A white dwarf is   c   a star which has run out of hydrogen fuel

4 A black hole is   d   hydrogen fusing to make helium

5 A star is   e   the massive remains of a supernova

6 A supernova is   f   a cooling star which sheds material into space [5]

Can-do tasks

Science is a practical subject and you deserve credit for being able to do practical things. Can-do tasks are an opportunity for you to demonstrate some of your practical and ICT skills throughout the course.

There are 81 of these tasks throughout your GCSE core Science course. Some are practical and some require the use of ICT.

The table summarises the number of tasks in Biology, Chemistry and Physics at each level.

	Biology	Chemistry	Physics
Level 1	11	11	9
Level 2	13	4	10
Level 3	6	8	9

You can only count a maximum of eight of these tasks and these tasks are set at three levels.

Level 1 (worth 1 mark): These are simple tasks that you can usually complete quickly. You may have done many of these before you started the course.

Level 2 (worth 2 marks): These are slightly harder tasks that might take a little longer to do.

Level 3 (worth 3 marks): These are even more difficult tasks that may take you some time to do.

Here are some examples of Can-do tasks:

Module and item	Can do task	Level
B1a	I can measure breathing rate/pulse rate before and after different types of exercise	1
C1h	I can measure the temperature of a liquid	1
P2f	I can use ICT to produce a labelled model of our Solar System	1
B1b	I can carry out simple food tests	2
C2d	I can extract a sample of copper from a copper ore such as malachite	2
P2b	I can use an oscilloscope to measure the maximum voltage of AC	2
B2a	I can investigate and compare different habitats	3
C2g	I can measure the rate of reaction that produces a gas	3
P1a	I can carry out an experiment to find the energy needed to melt ice	3

These are only a small selection of the Can-do tasks you can attempt. In the left hand column it tells you where this task might be attempted, for example B1b the food tests are in Biology Module 1, item b. Some of these tasks, such as measuring the temperature of a liquid, may turn up many times in the course in physics, chemistry or biology. They can be scored at any time not just in C1h.

Some of the Can-do tasks produce a product, for example the model of the Solar System. This can be very useful as evidence that you have successfully completed a task.

Your teacher has to see that you have completed a task and record this on a record sheet. Remember only the best eight can count so the maximum is $8 \times 3 = 24$.

Your teacher may give you a list of all the Can-do tasks at the start of the course. You can then tell him or her when you think you have successfully completed one of them. They can then give you credit for this.

Remember, it does not matter if you fail to do a task or if you are absent. There will be many more chances throughout the course.

These Can-do tasks must be completed in order to count. You cannot be awarded two marks, for example, if you nearly completed a three mark task.

If you go on to take the GCSE Additional Science, there are different Practical Activities to the ones in this GCSE. They are more appropriate for somebody going forward to do sciences beyond GCSE. However, the Can-do tasks that you do for GCSE Science can also count if you go on to do either separate Physics, Chemistry or Biology GCSEs.

Finally, just how important is a Can-do task? Every time you complete even a Level 3 task it is worth more to your final result than scoring 4 marks on a written paper.

Good luck!

Science in the News

Do you read a newspaper or listen to radio or television news programmes?

Do you believe everything you read or hear?

Two headlines in a national newspaper on 28th October 2005 were:

'Bird Flu Man is a Hospital Worker'

and

'Weekly Helping of Broccoli May Cut Lung Cancer Risk'

Looking at the first headline you might think the man had bird flu. But if you had read the article you would have discovered that he was the man who ran a quarantine centre where two parrots died and he also worked in a hospital. He never had bird flu.

Reading the second article would tell you that the sample used was so small that scientists could not be sure that a weekly helping of broccoli would cut the risk of getting cancer.

As part of your GCSE core Science you have to do at least one Science in the News task. If you do more than one, only your best mark will count. In the Science in the News task you have to use your knowledge of science to solve a problem.

The task will be in the form of a question, for example should smoking be banned in public places? With the question you will be given some 'stimulus' material to help you and approximately one week to do some research.

What should you do with the stimulus material?

Read it through carefully and identify any scientific words you do not understand. Look up the meaning of these words. The glossary in this book might be the starting point. Then go through with a highlighter pen and highlight those parts of the stimulus material you might want to use to answer the question.

What research should you do?

You should be looking for at least two or three sources of information. These could be from books, magazines, the Internet or CD-ROMs. You could also use surveys or experiments.

You will need to include with your report a list of sources that are detailed enough that somebody could check them. Some of your sources could look like this:

1 www.webelements.com/webelements/scholar/index.html

2 Heinemann, *Gateway Science: OCR Science for GCSE*, Higher text book p. 75–77

3 *Daily Mail*, 26th October 2005 p.2.

You can take this research material with you when you have to write your report about one week later. Do not print out vast amounts of irrelevant material from the Internet because you will not be able to find what you want when you write your report. Your teacher will probably collect in your research material to help them to assess your report but they will not actually mark it.

If you choose to do no research it does not stop you writing a report but you will get a lower mark.

Writing your report

You will have to write your report in a lesson supervised by the teacher. It has to be your own work. There is no time limit but if you need longer than the lesson allowed, all work must be collected in and stored securely until the next time.

You report should be between 400 and 800 words. As you write your report you need to refer to the information you have collected. For example, 'scientists believe that the long-term demand for electricity in the UK can only be met with a new generation of nuclear reactors (3)'. The 3 in brackets refers to source 3 from your table of sources, *Daily Mail*.

You should be critical of the sources. You will soon find that everything you read in newspapers is not necessarily true. Make sure you answer the question. When you finish, read your report through carefully. Your teacher may give you details of the criteria they are using to mark your report.

Marking your report

Your teacher will mark your report against simple criteria and look for six skills each marked out of a maximum of 6. This makes a total of 36. Your teacher will explain these criteria to you.

This report is worth about 20% of the total marks. Hopefully, writing this report will make you more aware of science in our everyday lives.

Useful data

Physical quantities and units

Physical quantity	Unit(s)
length	metre (m); kilometre (km); centimetre (cm); millimetre (mm)
mass	kilogram (kg); gram (g); milligram (mg); micrograms (µg)
time	second (s); millisecond (ms)
temperature	degree Celsius (°C); kelvin (K)
current	ampere (A); milliampere (mA)
voltage	volt (V); millivolt (mV)
area	cm^2; m^2
volume	cm^3; dm^3; m^3; litre (l); millilitre (ml)
density	kg/m^3; g/cm^3
force	newton (N)
speed	m/s; km/h
energy	joule (J); kilojoule (kJ); megajoule (MJ)
power	watt (W); kilowatt (kW); megawatt (MW)
frequency	hertz (Hz); kilohertz (kHz)
gravitational field strength	N/kg
radioactivity	becquerel (Bq)
acceleration	m/s^2; km/h^2

Electrical symbols

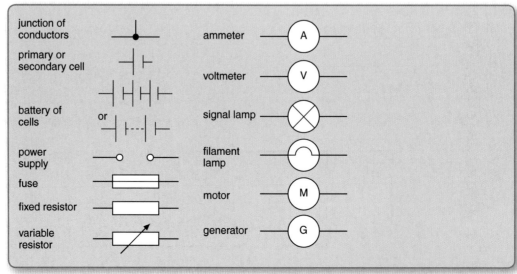

Periodic Table

Key

relative atomic mass
atomic symbol
name
atomic (proton) number

1	2

1
H
hydrogen
1

7	9
Li	**Be**
lithium	beryllium
3	4

23	24
Na	**Mg**
sodium	magnesium
11	12

39	40	45	48	51	52	55	56	59	59	64	65
K	**Ca**	**Sc**	**Ti**	**V**	**Cr**	**Mn**	**Fe**	**Co**	**Ni**	**Cu**	**Zn**
potassium	calcium	scandium	titanium	vanadium	chromium	manganese	iron	cobalt	nickel	copper	zinc
19	20	21	22	23	24	25	26	27	28	29	30

85	88	89	91	93	96	[98]	101	103	106	108	112
Rb	**Sr**	**Y**	**Zr**	**Nb**	**Mo**	**Tc**	**Ru**	**Rh**	**Pd**	**Ag**	**Cd**
rubidium	strontium	yttrium	zirconium	niobium	molybdenum	technetium	ruthenium	rhodium	palladium	silver	cadmium
37	38	39	40	41	42	43	44	45	46	47	48

133	137	139	178	181	184	186	190	192	195	197	201
Cs	**Ba**	**La***	**Hf**	**Ta**	**W**	**Re**	**Os**	**Ir**	**Pt**	**Au**	**Hg**
caesium	barium	lanthanum	hafnium	tantalum	tungsten	rhenium	osmium	iridium	platinum	gold	mercury
55	56	57	72	73	74	75	76	77	78	79	80

[223]	[226]	[227]	[261]	[262]	[266]	[264]	[267]	[268]	[271]	[272]
Fr	**Ra**	**Ac***	**Rf**	**Db**	**Sg**	**Bh**	**Hs**	**mt**	**Ds**	**Rg**
francium	radium	actinium	rutherfordium	dubnium	seaborgium	bohrium	hassium	meitnerium	darmstadtium	roentgenium
87	88	89	104	105	106	107	108	109	110	111

3	4	5	6	7	8
					4
					He
					helium
					2

11	12	14	16	19	20
B	**C**	**N**	**O**	**F**	**Ne**
boron	carbon	nitrogen	oxygen	fluorine	neon
5	6	7	8	9	10

27	28	31	32	35.5	40
Al	**Si**	**P**	**S**	**Cl**	**Ar**
aluminium	silicon	phosphorous	sulfur	chlorine	argon
13	14	15	16	17	18

70	73	75	79	80	84
Ga	**Ge**	**As**	**Se**	**Br**	**Kr**
gallium	germanium	arsenic	selenium	bromine	krypton
31	32	33	34	35	36

115	119	122	128	127	131
In	**Sn**	**Sb**	**Te**	**I**	**Xe**
indium	tin	antimony	tellurium	iodine	xenon
49	50	51	52	53	54

204	207	209	[209]	[210]	[222]
Tl	**Pb**	**Bi**	**Po**	**At**	**Rn**
thallium	lead	bismuth	polonium	astatine	radon
81	82	83	84	85	86

Elements with atomic numbers 112–116 have been reported but not fully authenticated

* The lanthanoids (atomic numbers 58–71) and the actinoids (atomic numbers 90-103) have been omitted.

Glossary

acid rain rain with a pH below about 6 formed when pollutants such as sulfur dioxide and nitrogen oxides dissolve

active immunity immunity developed by the body to foreign invading organisms

adapt the characteristics of an organism that make it well suited to living in a particular environment

addition polymerisation polymerisation where monomers join together to form a polymer without any loss of atoms

aerobic respiration respiration with oxygen

alkane a family of hydrocarbons containing only single carbon–carbon bonds. Alkanes have a general formula C_nH_{2n+2}

alkene a family of hydrocarbons containing a carbon–carbon double bond. Alkenes have a general formula C_nH_{2n}

alleles a genetic instruction received from one parent. Alleles from both parents form a gene

alloy a mixture of metals or a metal and carbon (in the case of steel)

alpha particle emitted by decay of some radioactive materials

alternating current (AC) flow of charge in a circuit, which keeps on changing direction

amplitude the distance of a crest of a wave from its rest position

anaerobic respiration respiration without oxygen

analogue signal that can have any value within a range

anode a positively charged electrode

antibiotic medicine that kills bacteria

antibody chemical produced by the body's immune system to destroy foreign invading organisms

antigen a thing that is foreign to the body

antioxidants substances that slow down the rate of oxidation of food

asteroids lumps of rock, smaller than a planet, in orbit around the Sun

atmospheric pollution contaminants of the environment that are a by-product of human activity. They include particles (smoke) and gases such as sulfur dioxide

axon the long extended part of a nerve cell

background radiation radiation from the surroundings

baking powder supplies carbon dioxide during the baking process so cakes will rise. It contains sodium hydrogen carbonate

bases the four chemicals A, C, T and G that code for the instructions for life in DNA

benign not malignant. Will not grow like a cancer

beta particle emitted by decay of some radioactive materials

Big Bang theory theory of the start of the Universe

bile liquid that emulsifies fats, produced in the liver and stored in the gall bladder

binocular vision using both eyes to view the same object so that distance can be judged more accurately

binomial naming organisms with two names, one for genus and one for species

biodegradable substances that can be broken down by such processes as decomposition by bacteria and can therefore be reused by living organisms.

biodiversity range of different kinds of living organisms

biomass plant material which can be burnt or fermented

black hole remnant of a supernova, so massive that light cannot escape

boiling point temperature at which a liquid rapidly turns to a vapour. Water has a boiling point of 100 °C at normal atmospheric pressure

carbohydrases enzymes that digest carbohydrates

carbohydrate a compound of carbon, hydrogen and oxygen which fits a formula $C_x(H_2O)_y$. Glucose, $C_6H_{12}O_6$, is an example of a carbohydrate

carbon dioxide a gas produced by respiration by both plants and animals and then used by plants for photosynthesis

carbon monoxide a compound of one carbon atom and one oxygen atom, CO. It is formed by the incomplete combustion of carbon and carbon compounds. Carbon monoxide is colourless, odourless and poisonous

catalyst a substance that alters the rate of a chemical reaction without being used up

cathode a negatively charged electrode

cavity walls walls that have two layers with a gap between them

cellulose substance made by plants that forms the structure of their cell walls

cement a substance made by mixing powdered limestone with clay. When mixed with water it sets to a hard mass

centripetal force force needed to keep an object moving along a circular path

chemical digestion breaking food down using enzymes

chromosome structure composed of DNA and found in the nucleus of cells

cilia small hair-like structures on the surface of cells

cirrhosis disease where the liver becomes damaged

climate change changes in the climate such as global warming, brought about by the activities of humans

colloid a state where very small particles of one substance are spread evenly through another

colour blindness not being able to see certain colours such as red and green

combustion burning of a substance with oxygen to release energy. Another word for burning

comets lumps of ice and dust in a highly elliptical orbit around the Sun

competition continual struggle that organisms have with each other for resources

complete combustion when a substance burns in a plentiful supply of air or oxygen to release the maximum amount of energy

composite a material that is made up of other materials

concentration quantity of solute dissolved in a stated volume of solvent

concrete a construction material using cement, sand and aggregate (small stones) mixed with water

conduction how heat energy is transferred through solids

construction materials materials that are used in building

contraception prevention of pregnancy

convection currents circulating movement of a heated fluid caused by differing densities

convection how heat energy is transferred by the bulk motion of liquids and gases

core the centre part of the Earth

corrosion the wearing away of the surface of a metal by chemical reaction with air and water

cosmic rays nuclear radiation from space

covalent type of bonding between atoms, usually non-metal atoms, involving a sharing of electrons.

cracking breaking down of long-chain hydrocarbon molecules by the action of a heated catalyst or by heat alone to produce smaller molecules

crater circular depression left in surface by impact of an asteroid

crest the highest point reached by a wave in a cycle

critical angle waves which hit a boundary at less than this angle will be completely reflected

cross-linking the bonds joining different monomer chains

crude oil a mixture of hydrocarbons produced by the action of high temperatures and pressures on the remains of sea creatures over millions of years

crust the outer layer of the Earth

Darwin the man who first put forward the theory of evolution

deforestation destruction of forests caused by excessive cutting down of trees

dehydration when an organism becomes short of water

dendrites cellular extensions of nerve cells

depressants drugs that depress neural activity

diabetes disease in which a person does not produce enough insulin to control the level of sugar in the blood

diastolic blood pressure when the heart is relaxing

diffraction spreading out of waves as they pass through a gap

diffusion a movement of particles from an area of high concentration to an area of low concentration

digital signal which can only have one of two values within a range

direct current (DC) current caused by a flow of charge in just one direction

DNA the molecules that code for the instructions to make a new living organism

dominant an allele that always expresses itself

drugs chemicals that produce a change in the body

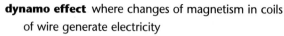
dynamo effect where changes of magnetism in coils of wire generate electricity

earthquake sudden release of energy in the Earth's crust, which generates waves

ecological niche part of a habitat

ecosystem a system of interacting organisms that live in a particular habitat

effector an organ such as a muscle that causes a response to a stimulus

efficiency ratio of the useful energy output to the total energy input in an energy transfer process

electrolysis the decomposition of a compound by the passage of electricity

electrolyte a compound split up by an electric current when molten or in solution

electron particle in an atom which is outside the nucleus

elliptical shaped like an egg, a squashed circle

emulsification the process of breaking down fat droplets into smaller ones

emulsifier (or emulsifying agent) a chemical that coats the surface of droplets of one liquid so they can remain dispersed in the other

emulsion mixture of two immiscible liquids where one liquid is dispersed in small droplets throughout the other

endangered organisms that are in danger of becoming extinct

endothermic a reaction that takes in energy from the surroundings

E-numbers system for listing permitted food additives. All permitted additives are given an E-number, e.g. E124

enzymes organic catalysts that speed up the rate of a reaction

ester a sweet-smelling liquid formed when an organic acid and an alcohol react. Esters are used in perfumes and food. Methyl ethanoate is an example of an ester

evolution adaptation of organisms to changes in the environment through natural selection

exothermic a reaction which gives out energy to the surroundings

explosion a very rapid reaction accompanied by a rapid release of gaseous products

exponentially an increase that becomes more rapid with time

extinct organisms that no longer exist on the Earth

fat food storage molecules made from fatty acids and glycerol

fertility the ability to fertilise when male and female sex cells fuse together

finite resource a resource whose life in limited and whose supply will run out in the future

food additive a substance added to food to act as a colouring agent, preservative, emulsifier, flavour enhancer, etc

fossil preserved remains or cast of a dead organism

fossil fuel fuel produced from the slow decay of dead animals and plants at high temperatures and high pressures

fraction product collected on fractional distillation of crude oil. A fraction has a particular boiling point range and particular uses

fractional distillation method for separating liquids with different boiling points

frequency the number of oscillations of a wave in one second

gamete cell involved in reproduction such as an ovum or a sperm

gamma high frequency wave emitted by decay of some radioactive materials

gene a section of DNA that codes for one specific instruction

generator device which transfers the kinetic energy of a rotating shaft into electrical energy

genetic code sequence of bases that code for the instructions to make a new living organism

glucose a type of sugar that is produced by photosynthesis and used as an energy source in respiration

granite an igneous rock formed inside the Earth by crystallisation of molten magma

hallucinogens mind altering drugs

heat energy energy required to change the temperature of an object

heat-stroke a condition caused when the body overheats

heterozygous a gene that consists of two different alleles

homeostasis maintaining a constant internal environment within the body

homozygous a gene that consists of two identical alleles

host a live organism that is fed upon by a parasite

hybrid a cross between two true breeding parents

hydrocarbon compound of carbon and hydrogen only

hydrophilic a substance that has a liking or attraction for water

hydrophobic a substance that repels water

hypothermia a lowering of the body's temperature

igneous rock formed when magma crystallises

incomplete combustion when substances burn in an insufficient amount of air or oxygen. It results in less energy release and possibly soot and/or carbon monoxide

indicator species the presence of a species that indicates the quality of the habitat

infrared a wave which transfers heat energy. Region of the electromagnetic spectrum between visible light and microwaves

insoluble describes a substance that does not dissolve in a solvent

insulin hormone produced by the pancreas that lowers the level of glucose in the blood

intermolecular forces between different molecules

invertebrates animals without a backbone

ionisation removal from or addition of electrons to a particle

ionosphere layer of charged particles high in the Earth's atmosphere

key a means of identifying different organisms

kilowatt one thousand watts

kilowatt-hour the energy transferred by an electrical device with a power of one kilowatt over an hour

kinetic energy energy transferred to an object to set it in motion

kwashiorkor disease caused by a lack of protein in the diet

lactic acid chemical that causes muscle fatigue and is produced during anaerobic respiration

laser device which emits electromagnetic waves, all with the same frequency and direction

lava molten rock that escapes from a volcano

light-year distance travelled by light passing through space for one year

limestone a sedimentary rock formed from the remains of sea creatures. It is a form of calcium carbonate

limiting factor a factor such as light, temperature or carbon dioxide where a lack of it limits the rate of photosynthesis

lipases enzymes that digest fat

lithosphere the crust and the uppermost layer of the mantle

long sight when the eye focuses light behind the retina

longitudinal type of wave which produces oscillations parallel to its direction of travel

LPG (liquefied petroleum gas) the gas which leaves the top of the fractional distillation column and is liquefied. It is an alternative to petrol or diesel in cars

magma rock between the crust and the core of the Earth

magnetic field lines of force around an electric current which affect magnetic materials

malignant tumour that consists of rapidly dividing cells and is cancerous

mantle a thick layer of dense semi-liquid rock below the Earth's crust

marble a metamorphic rock formed by the action of high temperatures and pressures on limestone. A form of calcium carbonate

microwave region of the electromagnetic spectrum which lies between infrared and radio waves

monomer small molecule that joins together with other molecules to produce a polymer

Morse code alphanumeric code made from long and short pulses

motor a nerve cell that carries instructions away from the brain

mule sterile cross between a male donkey and a female horse

multiplexing method of sending many messages down a single link simultaneously

mutation a change to the structure of a gene or DNA caused by such things as chemicals, X-rays or radiation

mutualism a relationship between two organisms of different species in which both organisms benefit

national grid cables that carry electrical energy from power station to consumer

natural selection a process in which organisms that are most suited to the environment survive and produce more offspring

Near-Earth Object comet or asteroid whose orbit crosses that of the Earth

neurone a name for a nerve cell

neutron star very dense remnant of a supernova, not heavy enough to form a black hole

nicotine an addictive chemical that is found in tobacco

non-renewable fuel a fuel that took a long time to form and cannot quickly be replaced

nuclear power energy source which uses radioactive fuel

nuclear radiation emission of ionising radiation from changes in a nucleus

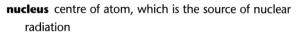

nucleus centre of atom, which is the source of nuclear radiation

oestrogen a hormone produced by the ovary that causes the lining of the uterus to thicken

oil liquid fat

optical fibre thin strand of very transparent glass for carrying pulses of light long distances

orbit path followed by an object bound by the gravity of another object

ore a rock that contains a metal or a metal compound in sufficient quantity to make the extraction of the metal economically viable

organism a living animal or plant

ovulation relase of an ovum from the ovary

oxygen debt caused by anaerobic respiration and the production of lactic acid. After exercise the debt has to be repaid to break down the lactic acid

ozone a form of oxygen which can absorb ultraviolet light from the Sun preventing it reaching the Earth

pain killers drugs that provide relief from pain

parasite organism that lives on or in another living organism causing it harm

passive immunity short lasting immunity that is gained by injecting other peoples' antibodies

pathogens disease-causing organisms

payback time time taken before the money saved in heating equals the installation cost

performance enhancers drugs that increase performance

phosphorescent pigments pigments that store energy when in light and can release this energy again in the dark and so glow in the dark

photocell device that transfers light energy into electrical energy

photosynthesis the process by which plants convert water and carbon dioxide into oxygen and glucose using the energy from the Sun

planet object which orbits around a star

planetary nebula a dying star which puffs out material into space

plutonium waste product of nuclear power, which can be used to make atomic bombs

pollution the presence in the environment of substances that are harmful to living things

polymer long chain molecule built up of a large number of smaller units, called monomers, joined together by the process of polymerisation

population a group of organisms of one species living together in a habitat

power the energy transferred by a device in one second

predator an animal that hunts and kills other animals for food

prey an animal that is hunted and killed for food by a predator

products substances that are formed in a chemical reaction. They appear to the right of the arrow in a chemical equation

progesterone a hormone produced by the ovary after the ovum has been released. It maintains the wall of the uterus during pregnancy

proteases enzymes for digesting protein

protein large polymer molecule made from a combination of amino acids

P-wave longitudinal wave produced by an earthquake

quadrat a square grid used to determine population density

radiation how heat energy is transferred by infrared waves

radioactive materials substances which emit nuclear radiation

radioactive waste material left over when uranium has been used to make electricity

rate of reaction the speed with which products are formed or reactants used up

reactants substances that are used at the start of a chemical reaction. They appear to the left of the arrow in a chemical equation

receptor an organ or cell that receives an external stimulus

recessive an allele that only expresses itself if the dominant allele is not present

recycled materials that are used again rather than disposed of

red giant star which is fusing helium to form carbon

reflex arc the pathway along which nerve impulses pass in a simple reflex action

refraction the change in direction of a wave as it passes from one material to another

relay a nerve cell that passes an impulse between two other nerve cells

resources chemicals and materials that can be used for the benefit of humans

respiration a process that takes place in living cells converting glucose and oxygen into water and carbon dioxide with a release of energy

rust product formed when iron or steel are exposed to air and water. This brown compound is a hydrated iron(III) oxide

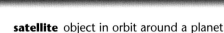

satellite object in orbit around a planet

saturated organic compounds where all bonds are single covalent bonds

sensory a nerve cell that carries information to the brain

sex hormones hormones that control the secondary sexual characteristics

sheath the fatty coat that surrounds a nerve cell

short sight when the eye focuses light short of the retina

smart alloy an alloy which can be restored to its original shape if deformed

solar flares clouds of charged particles ejected from the Sun

solute the substance that dissolves in a solvent to form a solution

solution what is formed when a solute dissolves in a solvent

solvent a liquid in which a solute dissolves

spacecraft device which can travel in space, away from Earth

species a group of organisms that breed and produce fertile offspring

specific heat capacity energy required to raise the temperature of one kilogram by one degree

specific latent heat energy required to change the state of a one kilogram of a material

star sphere of hot gas, which emits light and heat

starch a carbohydrate food storage substance produced by plants

state whether a substance is solid, liquid or gas

stimulants drugs that stimulate neural activity

subduction dipping of one plate below another at a destructive plate boundary

sun protection factor the extra time a cream allows you to stay in the Sun before burning

supernova explosion of a star when it becomes unstable at the end of its life

surface area the area of the surface of a solid object, usually measured in centimetres squared

sustainable development development using the Earth's resources at a rate at which they can be replaced

S-wave transverse wave produced by an earthquake

synapse a minute gap between two neurones

systolic blood pressure when the heart is contracting

tectonic plates very large plates of rock which float and move very slowly on the mantle of the Earth

temperature a measure of how hot or cold an object is

thermal decomposition decomposition of a compound by heating

thermochromic pigments pigments that can change colour at different temperatures

thermogram photograph taken in infrared light

total energy input the energy transferred into a device for it to accomplish its task

total internal reflection reflection inside a material when the angle of incidence exceeds the critical angle

toxicity a measure of how poisonous a substance is

toxins poisons produced by microorganisms

transformer device which raises or lowers the voltage of an alternating current

transmission frequencies range of waves used to carry information

transmitter a chemical that transmits a nerve impulse across a synapse

transverse waves type of wave which produces oscillations at right angles to its direction of travel

trough the lowest point reached by a wave in a cycle

tumours a group of unspecialised cells that may be malignant or benign

ultraviolet region of the electromagnetic spectrum between visible and X-rays

unsaturated organic compound containing one or more double or triple bond between carbon atoms

uranium can be used to transfer nuclear energy into heat energy

useful energy output the energy transferred by a device to do some task

vasoconstriction a decrease in the diameter of blood capillaries

vasodilation an increase in the diameter of blood capillaries

vector organism that transmits and carries a disease

vertebrates animals that have a backbone

volcano a tube from inside the Earth to its surface through which lava can escape in an eruption

watt unit of power

wavelength distance between adjacent crests (or troughs) along a wave

white dwarf remnant of a star which has run out of fuel for fusion reactions

wind turbine device which transfers kinetic energy of wind into electrical energy

wireless technology stuff which allows information to be transferred through space without wires

Index

acceleration, units 222
accommodation, of the eye 15
acid rain 65, 128, 224
active immunity 13, 224
adaptation 55–58, 224
addition polymerisation 93, 224
aerobic respiration 4, 224
air 131, 132
alcohol 19, 21
alcohol units in drinks 21
alkanes 92, 224
alkenes 92, 224
alleles 32, 224
alleles, dominant 32
alleles, recessive 32
alloys 123, 125, 224
alloys, smart 125
alpha particles 196, 224
alternating current (AC) 188, 224
aluminium 127, 128, 129
amalgam (mercury) 125
ammonia 133
ammonium nitrate 139
ampere 222
amphibians 44
amplitude of a wave 172, 224
anaerobic respiration 4, 224
analogue signals 163, 224
animals 44
anode 124, 224
antibiotics 13, 61, 224
antibodies 12, 224
antigens 12, 224
antioxidants 224
antioxidants in foods 79, 81
apes, evolution of 45
area, units 222
argon 132
artificial ecosystems 39, 68
aspirin 19
asteroids 224
asteroids, impacting the Earth 208
asteroids, near-miss 210
asteroids, past collisions 208–209
atmosphere 133
atmospheric pollution 131–134, 224
Aurora Borealis 200
axons 16, 17, 224

background radiation 195, 224

baking powder 77, 224
balanced diets 7
basalt 121
bases of DNA 28, 224
becquerel 222
benign growths 11, 224
beta particles 196, 224
Big Bang theory 212, 224
bile 9, 224
binding medium for paints 112
binocular vision 15, 224
binomial system of classification 45, 224
biodegradable 224
biodegradable polymers 97
biodiesel 194
biodiversity 39, 224
biomass 224
birds 44
black holes 213, 224
black smokers 58
blind drug testing 14
blindness 18
blood pressure 3
blood pressure, high 6
blood sugar 25
blue-shifted light 211
body image 10
body mass index (BMI) 10
boiling point 88, 224
bond breaking 88, 105
bond making 105
bonds, covalent 88
brass 125
breeding for characteristics 32
bronchitis 20
Bunsen burner flames 101
butane 88

cacti 57
cakes 77
calcium carbonate 116, 137
calcium hydroxide 104, 118
calcium oxide 116
calorimeters 104
camels 56
cancer 11
Can-do tasks 218–219
cannabis 22
carbohydrase enzymes 9, 224
carbohydrates 8, 9, 77, 224
carbon cycle 132
carbon dioxide 4, 5, 47, 48, 50, 132, 224

carbon dioxide, as a greenhouse gas 64
carbon dioxide, emissions 64
carbon dioxide, testing for 77, 101
carbon monoxide 20, 101, 131, 133, 224
carbon monoxide, poisoning from 102
cars, recycling 127, 128, 130
catalysts 140, 225
catalytic converters 133
cathode 124, 225
cavity wall insulation 157
cavity walls 225
cellulose 48, 225
Celsius (temperature scale) 147, 222
cement 117, 225
centripetal force 204, 225
Cepheid variable stars 214
CFCs 175
chemical digestion 9, 225
chloroethane 93
chromosomes 28, 225
chromosomes, genetic variety 29
cilia 20, 225
cirrhosis of the liver 21, 225
classification of drugs 20
climate change 67, 176, 225, 178
coal 100
colloids 112, 225
colour blindness 16, 225
colour vision 16
combustion 100, 132, 225
combustion, bond breaking and bond making 105
combustion, energy changes 104–105
combustion, energy transfer 105
comets 209, 225
competition for resources 51, 52, 225
complete combustion 100, 225
composite materials 117, 225
concave lenses 16
concentration 136, 225
concrete 115, 117, 225
conduction of heat 156, 225
conductors of heat 152, 156
conservation programmes 68–69

construction materials 115–118, 225
contraceptive pills 25, 225
convection 152, 156, 225
convection currents 225
convection currents within the Earth 120
convex lenses 16
cooking of food 76, 135
copper 123, 124, 126, 127, 129, 138
copper(II) sulfate 124
core of the Earth 120, 225
cornea 15
coronal mass ejection (CME) 202
corrosion 128, 225
cosmetics, testing 85
cosmic rays 195, 199, 200, 225
covalent bonds 88, 105, 225
cracking of crude oil 89, 225
craters 207, 225
crest of a wave 172, 225
critical angle 165, 225
crop growing 39
cross-linking of polymer chains 96, 225
crude oil 87–90, 225
crude oil, environmental problems 89
crude oil, political considerations 90
crust of the Earth 120, 225
current, alternating (AC) 188
current, direct (DC) 184
current, units 222

Darwin, Charles 60, 62, 225
defences against diseases 12
deforestation 225
dehydration 24, 225
dendrites 16, 17, 225
density, units 222
depressant drugs 19, 225
diabetes 18, 25, 225
diastolic blood pressure 3, 225
diet 7–10
diffraction of electromagnetic waves 169, 225
diffusion of nutrients 9, 225
digestion 9
digital radio 170
digital signals 163, 164, 225
dinosaurs, extinction of 55

direct current (DC) 184, 188, 225
diseases 11–14
diseases, inherited 33
DNA 27, 28, 225
DNA, mutations 31
dolphins 45
dominant alleles 32, 225
Doppler shift 211
double blind drug testing 14
drink-driving 21
drinking alcohol 21
drugs 226
drugs (medicines) and safety 14
drugs of addiction 19–22
dyes, for fabrics 111
dynamite 139
dynamo effect 188, 226

Earth, extraterrestrial threats 207–210
Earth, magnetic field 199–202
Earth, resources 63
Earth, structure 119, 120
earthquakes 119, 120, 177, 226
eco-friendly gloss paints 114
ecological niche 52, 226
ecology 39–42
ecosystems 39, 226
effectors 17, 226
efficiency 226
eggs, cooking 76, 78
electrical symbols 222
electricity generation 187–190
electricity generation, fuels 191–194
electricity generation, generators 188
electricity generation, methane fuel 191
electricity generation, nuclear power 191, 192, 198
electricity meters 193
electrolysis 226
electrolysis of copper 124, 126
electrolytes 124, 226
electromagnetic spectrum, wave properties 172
electron 226
elliptical 226
elliptcal orbits 209
emulsification 9, 226
emulsifiers 226
emulsifiers in foods 79, 80
emulsion paints 112
emulsions 80, 226
endangered species 67, 226
endangered species, helping 68
endothermic reactions 103, 226

endothermic reactions, recognising 104
energy efficiency 153
energy flow 148
energy in chemical reactions 103–106
energy release 4
energy sources and production 183, 187–190
energy sources and production, solar power 183–185, 186
energy sources and production, wind power 185, 186
energy, heat energy 147
energy, of waves 160–161
energy, sunlight 183–185, 186
energy, units 222
E-numbers in foods 80, 226
enzymes 226
enzymes for digestion 9
esters 83, 84, 226
ethane 92
ethanol, combustion 104
ethene 92
ethene, polymerisation 93
ethyl methanoate 84
evolution 61, 226
evolution, theories of 62
exothermic reactions 103, 106, 226
exothermic reactions, recognising 104, 105
explosions 139, 226
exponential growth 63
exponentially 226
extinction 67, 226
eyes 15
eyes, testing 18

fabric dyes 111
fatigue 5
fats 8, 9, 48, 226
fertility 25, 226
finite fuel resources 87
finite resources 226
fireworks 138
fish 44
fitness 3–6
flavour enhancers in foods 79
food additives 79–82, 226
food colours and dyes 79
food packaging 81, 82
food production, increasing 50
foods 8, 75
force, units 222
fossil fuels 87, 99, 191, 226
fossils 59, 226
fraction 226
fractional distillation 88, 226
fractions of crude oil 88

frequency, units 222
frequency of a wave 172, 226
fuels 89, 99–102

gabbro 121
galaxies 212
gametes 29, 226
gamma rays 196, 226
gamma rays, medical scanning 197
gas (fuel) 100
generators, of electricity 188, 226
genes 28, 226
genes, or environment 33
genes, switching on or off 28
genetic code 28, 226
genetics 27–34
glass 127
glass, critical angle 165
glaucoma 18
global warming 64
gloss paints 112
glucose 4, 47, 48, 226
Gore-Tex® 96
granite 116, 121, 226
gravitional field strength, units 222
greenhouse effect 64

habitats 52
habitats, damage to 67
habitats, protecting 68
hallucinogenic drugs 19, 226
harlequin ladybirds 54
heart 3
heat and heating 147–158
heat and heating, cost of heating 152, 153
heat and heating, energy efficiency 153, 154
heat and heating, heat flow 155
heat and heating, insulation 151–158
heat and heating, specific heat capacity 148
heat and heating, specific latent heat 149
heat energy 226
heat pumps 150
heat stroke 24, 226
heroin 19
hertz 222
heterozygous 32, 226
high blood pressure and health 6
homeostasis 23, 226
homozygous 32, 226
horses, evolution of 60
hosts to parasites 12, 53, 226
Human Genome Project 30
hunting whales 67

hybrids 43, 226
hydrocarbons 88, 92, 226
hydrogen peroxide 140
hydrogen, star formation 213
hydrophilic 80, 226
hydrophobic 80, 227
hydrothermal vents 58
hypothermia 24, 227

ice 148
igneous rocks 116, 121, 227
immunity 12–13
incomplete combustion 101, 227
indicator species 65, 66, 227
infrared radiation 156, 227
infrared radiation, communications 163–164
infrared radiation, cooking 160
inheritance, knowing about our genes 34
inherited diseases 33
insect pollination 57
insolubility 84, 85, 227
insulation 151–152, 155, 157
insulin 25, 227
intermolecular forces 96, 227
invertebrates 44, 227
ionisation 196, 227
ionosphere 168, 227
iron 127, 129, 140
iron, rusting of 128, 130

joule (unit of energy) 147, 222

kelvin 222
key for organism characterisation 40, 227
kilogram 222
kilowatt 227
kilowatt hours (kWh) 193, 227
kinetic energy 227
kinetic energy of moving air 185
kwashiorkor 8, 227

lactic acid 4, 5, 227
ladybirds 54
Lamarck, Jean Baptiste de 62
lasers 173, 227
lava 121, 227
lead 125
length, units 222
lens of the eye 15
lichens 66
life in low gravity 206
life on other planets 203
light 171–174
light, wave properties 172
light-year (unit of length) 205, 227
lime 116

limestone 115, 116, 117, 118, 227
limewater 77, 101
limiting factors 49, 227
linseed oil 112
lipases 9, 227
liquefied petroleum gases (LPG) 88, 227
lithium 138
lithosphere 120, 227
liver 5, 9
liver cirrhosis 21
long sight 16, 227
longitudinal 227
longitudinal waves 177
lungs and smoking 20

magma 121, 227
magnesium 138
magnetic field 227
magnetic field of the Earth 199–202
magnetism and electricity 188
maintaining body conditions 23–26
malachite (copper ore) 124
malaria 12
malignant tumours 11, 227
mammals 44
manganese(IV) oxide 140
manned space travel 205, 206
mantle 120, 227
marble 116, 227
Mars, measurments from robot spacecraft 174
Mars, mission to 158, 206
Mars Orbital Laser Altimeter (MOLA) 174
mascara 86
mass, units 222
meat 76
medicines (drugs) and safety 14
mercury 125
metal ores 115, 123
metals 123–126
metals, corrosion 129
meteorites 55
methane 92, 133, 191
methane, complete combustion 100–101
methane, incomplete combustion 101
metre 222
microorganisms and disease 12
microwave ovens 159, 160
microwaves 159, 160, 227
microwaves, communications 161, 162
microwaves, support for Big Bang theory 212
mobile telephones 167

mobile telephones, safety 162
monomers 91, 227
Moon 204
Moon, origin of 201
Moon, surface features 207
Morse code 171, 227
motor neurones 16, 227
MRSA superbug 13
mules 43, 227
multiplexing 164, 227
mutations 31, 227
mutualism 53, 227

national grid 187, 227
natural dyes 111
natural selection 60, 227
Near-Earth Objects (NEOs) 209, 210, 227
negative feedback 26
nervous system 16
neurones 16, 17, 227
neurotransmitters 17
neutron stars 213, 227
newspapers, science stories 220
newton (unit of force) 222
nickel/rhodium alloy 140
nicotine 20, 227
nitrogen 131, 132
nitrogen, oxides of 131, 133, 134
non-biodegradable polymers 97
non-renewable fuels 87, 227
nuclear power 191, 192, 198, 227
nuclear radiation 195–198, 227
nuclear radiation, medical scanning 197
nucleus 228
nylon 96

octane 88
oestrogen 25, 228
oil 228
oil (fuel) 100
oil-based paints 112
oils (edible) 48
OPEC 90
optical fibres 164, 165, 166, 228
optic nerve 15
orbits 204, 228
orbits, elliptical 209
ores 115, 123, 228
organisms 40, 228
organisms, counting 40–41
organisms, grouping 43–46
ovaries 25
ovulation 25, 228
oxidation 128

oxides of nitrogen 131, 133, 134
oxygen 4, 131, 132
oxygen debt 5, 228
oxygen scavengers 81
ozone 134, 228
ozone depletion 64
ozone layer 175

packaging for foods 81, 82
pain-killing drugs 19, 228
paints and pigments 111–114
pancreas 25
Pangea 122
parasites 12, 53, 228
passive immunity 13, 228
pathogens 12, 228
payback time for insulation 153, 228
peppered moths 60–61
performance-enhancing drugs 19, 228
perfumes and scents 83–86
perfumes and scents, testing 85
Periodic Table of elements 223
petrochemicals 87–90
phosphorescent pigments 113, 228
photocells 184, 228
photocells, pros and cons 184
photochemical smog 134
photosynthesis 47, 48, 50, 133, 228
photosynthesis, and respiration 49
physical quantities 222
pitch 87
placebos 14
planetary nebulae 213, 228
planets 204, 228
planets, orbits 204
plants 44, 47–50
plastics 95
plate tectonics, theory of 121, 122
platinum 140
plutonium 192, 228
polar bears 56
pollination of plants 57
pollution 63, 64, 131–134, 228
poly(ethene) 91, 94
polymerisation 91, 93
polymers 89, 91–98, 127, 228
polymers, disposal 97, 98
polymers, structure 96
populations 40, 51–54, 228
potassium 138
power 193, 228
power stations 189
power stations, cost to consumers 193

power stations, nuclear power 191, 192, 198, 227
power, units 222
predators 52, 228
pressure cookers 135
prey 52, 228
products of reactions 137, 228
progesterone 25, 228
propane 92
propene 92
proteases 9, 228
proteins 8, 9, 48, 76, 228
punnet squares 33
P-waves 177, 228

quadrats 41, 228
quicklime 116

radiation 228
radiation, infrared 156, 160, 163–164
radiation, light 171–174
radiation, microwaves 159, 160
radiation, nuclear 195–198
radiation, of heat 152, 156
radiation, ultraviolet (UV) 175
radioactive materials 195, 228
radioactive waste 197, 198, 228
radioactivity, units 222
radio reception 169
radio reception, digital radio 170
radiotherapy 197
radon 195
rates of reaction 135–142, 228
rates of reaction, increasing 136, 141, 137
rats 61
reactants 136, 228
reactivity series 130
receptors 17, 228
recessive alleles 32, 228
recycled 228
recycling 123, 127, 128, 130
recycling of polymers 97
red giant stars 213, 228
red-shifted light 211
reflection of electromagnetic waves 168
reflex actions 17
reflex arc 17, 228
refraction 165, 228
refraction of electromagnetic waves 168, 170
relay neurones 17, 228
remote controls 163–164
renewable biomass 191
report writing 221
reproduction, sexual 29
reptiles 44
research resources 220–221

resources of the Earth 228
respiration 48, 132, 228
respiration, and photosynthesis 49
retina 15
rhyolite 121
robots in space 205
rocks, igneous 116, 121
rust 228
rusting 128, 130

satellite communications 168
satellites 201, 228
saturated compounds 92, 229
scents and perfumes 83–86
scents and perfumes, testing 85
science in the news 220
second 222
seismometers 177
senses 15–18
sensory neurones 16, 229
sex hormones 24–25, 229
sex inheritance 33
sexual reproduction 29
shape-memory alloys (SMA) 125
sharks 45
sheath, of neurones 17, 229
shivering 24
short sight 15, 16, 229
sight testing 18
skin, temperature control 24
slaked lime 118
smart alloys 125, 229
smoke detectors 196
smoker's cough 20
smoking 19, 20
sodium 138
sodium hydrogencarbonate 77
solar energy 183–185, 186
solar flares 201, 229
solar flares, power blackouts 202
Solar System 203–206
solder 125
solubility 85
solutes 85, 229
solutions 85, 229
solvents 83, 85, 112, 229
spacecraft 174, 205, 206, 229
space travel 206
species 39, 45, 229
specific heat capacity 148, 229
specific latent heat 149, 229
speed, units 222
speed of a wave 172
spinal cord 17
stainless steel 128
starch 48, 97, 229
stars 204, 229
stars, life and death of 211–214

states of matter 149, 229
steam 148–149
steel 127, 128
stimulant drugs 19, 229
storage heaters 154
subduction 121, 229
sulfur dioxide 65, 131
Sun 204
sun protection factor (SPF) 176, 229
sunscreen lotions 176
superbugs 61
supernovae 213, 229
surface area 229
surface area and reaction rates 141
sustainable development 69, 229
sustainable resources 69
S-waves 177, 229
sweating 24
synapses 17, 19, 229
synthetic dyes 111
systolic blood pressure 3, 229

tablets, rates of dissolving 142
tapeworms 53
'TEACUPS', and fuel properties 100
tectonic plates 120–121, 122, 229
telephones, mobile 162, 167
temperature 147, 229
temperature, units 222
temperature control 26
temperature control of bodies 24
testes 25
testing of drugs 14
thermal decomposition 116, 229
thermochromic paints and pigments 113, 229
thermograms 147, 151, 229
time, units 222
tin 125
titanium(IV) chloride 140
total energy input 229
total internal reflection (TIR) 164, 165, 229
toxicity 229
toxins 12, 229
transformers 187, 229
transmission frequencies for radio 169, 170, 229
transmitters 229
transverse waves 177, 229
trinitrotoluene (TNT) 139
trough of a wave 172, 229
tumours 11

ultraviolet radiation (UV) 175, 229

ultraviolet radiation (UV), sun protection 176
understanding science stories in the news 220
units of alcohol in drinks 21
units of electricity consumption 193
units used for scientific measurements 222
Universe, age of 214
Universe, expanding 211–212
Universe, origin 212
unsaturated compounds 92, 229
unsaturated compounds, testing for 93
uranium 191, 192, 229
useful energy output 229

vaccines 13
vanadium(V) oxide 140
vasoconstriction 229
vasodilation 229
vectors of diseases 12, 229
vegans 7
vegetarians 7
vertebrates 44, 229
vibrations as heat carriers 157
vision 15
vision, testing 18
volcanoes 119, 121, 176, 229
volt (V) 222
voltage, units 222
volume, units 222

warfarin 61
water, boiling point 147
water, ice and steam 148
water-based paints 112
watt (W) 193, 222, 229
wavelength 172, 229
waves 172
waves, calculating the speed 172–173
waves, diffraction 169
waves, energy 160–161
waves, longitudinal 177
waves, reflection 168
waves, refraction 168, 170
waves, transverse 177
whale watching 70
white dwarf stars 213, 229
wind pollination 57
wind power 185, 186
wind turbines 185, 229
wireless technology 167–170, 229
wireless technology, infrared radiation 163–164
wireless technology, microwaves 161, 162
writing reports 221

X chromosomes 33

Y chromosomes 33

zinc 125, 129, 130

Revision Guides

Beat the rest - exam success with Heinemann

Ideal for homework and revision exercises, these differentiated **Revision Guides** contain everything needed for exam success.

- Summary of each item at the start of each section

- Personalised learning activities enable students to review what they have learnt

- Advise from examiners on common pitfalls and how to avoid them

Please quote S 603 SCI A when ordering

(t) 01865 888068   (f) 01865 314029   (e) orders@heinemann.co.uk   (w) www.heinemann.co.uk

Inspiring generations

L554